CLIMATE CHANGE

NOMOS

LXVII

NOMOS

Harvard University Press
I *Authority* 1958, reissued in 1982 by Greenwood Press

The Liberal Arts Press
II *Community* 1959
III *Responsibility* 1960

Atherton Press
IV *Liberty* 1962
V *The Public Interest* 1962
VI *Justice* 1963, reissued in 1974
VII *Rational Decision* 1964
VIII *Revolution* 1966
IX *Equality* 1967
X *Representation* 1968
XI *Voluntary Associations* 1969
XII *Political and Legal Obligation* 1970
XIII *Privacy* 1971

Aldine-Atherton Press
XIV *Coercion* 1972

Lieber-Atherton Press
XV *The Limits of Law* 1974
XVI *Participation in Politics* 1975

New York University Press
XVII *Human Nature in Politics* 1977
XVIII *Due Process* 1977
XIX *Anarchism* 1978
XX *Constitutionalism* 1979
XXI *Compromise in Ethics, Law, and Politics* 1979
XXII *Property* 1980
XXIII *Human Rights* 1981
XXIV *Ethics, Economics, and the Law* 1982
XXV *Liberal Democracy* 1983
XXVI *Marxism* 1983
XXVII *Criminal Justice* 1985

XXVIII *Justification* 1985
XXIX *Authority Revisited* 1987
XXX *Religion, Morality, and the Law* 1988
XXXI *Markets and Justice* 1989
XXXII *Majorities and Minorities* 1990
XXXIII *Compensatory Justice* 1991
XXXIV *Virtue* 1992
XXXV *Democratic Community* 1993
XXXVI *The Rule of Law* 1994
XXXVII *Theory and Practice* 1995
XXXVIII *Political Order* 1996
XXXIX *Ethnicity and Group Rights* 1997
XL *Integrity and Conscience* 1998
XLI *Global Justice* 1999
XLII *Designing Democratic Institutions* 2000
XLIII *Moral and Political Education* 2001
XLIV *Child, Family, and State* 2002
XLV *Secession and Self-Determination* 2003
XLVI *Political Exclusion and Domination* 2004
XLVII *Humanitarian Intervention* 2005
XLVIII *Toleration and Its Limits* 2008
XLIX *Moral Universalism and Pluralism* 2008
L *Getting to the Rule of Law* 2011
LI *Transitional Justice* 2012
LII *Evolution and Morality* 2012
LIII *Passions and Emotions* 2012
LIV *Loyalty* 2013
LV *Federalism and Subsidiarity* 2014
LVI *American Conservatism* 2016
LVII *Immigration, Emigration, and Migration* 2017
LVIII *Wealth* 2017
LIX *Compromise* 2018
LX *Privatization* 2018
LXI *Political Legitimacy* 2019
LXII *Protest and Dissent* 2020
LXIII *Democratic Failure* 2020
LXIV *Truth and Evidence* 2021
LXV *Reconciliation and Repair* 2023
LXVI *Civic Education in Polarized Times* 2024
LXVII *Climate Change* 2026

NOMOS LXVII
Yearbook of the American Society for Political and Legal Philosophy

CLIMATE CHANGE

Edited by
Chiara Cordelli and Melissa Lane

New York University Press • *New York*

NEW YORK UNIVERSITY PRESS
New York
www.nyupress.org

© 2026 by New York University
All rights reserved

Please contact the Library of Congress for Cataloging-in-Publication data.

ISBN: 9781479842094 (hardback)
ISBN: 9781479842124 (library ebook)
ISBN: 9781479842100 (consumer ebook)

This book is printed on acid-free paper, and its binding materials are chosen for strength and durability. We strive to use environmentally responsible suppliers and materials to the greatest extent possible in publishing our books.

The manufacturer's authorized representative in the EU for product safety is Mare Nostrum Group B.V., Mauritskade 21D, 1091 GC Amsterdam, The Netherlands. Email: gpsr@mare-nostrum.co.uk.

Manufactured in the United States of America

10 9 8 7 6 5 4 3 2 1

Also available as an ebook

CONTENTS

Contributors — ix

Introduction — 1
CHIARA CORDELLI AND MELISSA LANE

PART I: THE LIMITS OF CLIMATE ECONOMICS

1. Ways Not to Think About Climate Change — 13
   DOUGLAS A. KYSAR

2. Climate Change, Inequality, and Expert Knowledge — 66
   ZEYNEP PAMUK

PART II: CLIMATE CHANGE AND THE CAPITALIST ORDER

3. Domination in the Age of the Externality — 79
   ALYSSA BATTISTONI

4. Environmental Justice, Capitalism, Democracy — 125
   MARK BUDOLFSON

5. The Chicago School's Coasean Incoherence — 141
   MADISON CONDON

PART III: CLIMATE CHANGE AND MORAL RESPONSIBILITY

6. On the Moral Challenge of the Climate Crisis — 165
   LUCAS STANCZYK

7. Nonconsequentialism and Climate Change — 199
   F. M. KAMM

8. Kicking Cans and Taking Stock 222
   STEVE VANDERHEIDEN

9. Hard Truths in Climate Policy and Politics 242
   SHELLEY WELTON

10. The Environmental Argument for Immigration
    Restrictions: A Critique 264
    JAMIE DRAPER

    Index 295

CONTRIBUTORS

Alyssa Battistoni
Assistant Professor of Political Science, Barnard College

Mark Budolfson
Associate Professor of Philosophy, and of Geography and the Environment, University of Texas at Austin

Madison Condon
Associate Professor of Law, Boston University School of Law

Chiara Cordelli
Professor of Political Science at the University of Chicago, and Senior Research Fellow at the Centre for History and Economics in Paris at Sciences Po

Jamie Draper
Assistant Professor of Philosophy, Utrecht University

Douglas A. Kysar
Joseph M. Field '55 Professor of Law, Yale University

F. M. Kamm
Henry Rutgers University Professor of Philosophy and Distinguished Professor of Philosophy, Rutgers University, and Littauer Professor of Philosophy and Public Policy Emerita, Harvard University

Melissa Lane
Class of 1943 Professor of Politics, Princeton University

Zeynep Pamuk
Associate Professor of Contemporary Political Theory and Professorial Fellow, Nuffield College, University of Oxford

Lucas Stanczyk
Associate Professor of Philosophy, Harvard University

Steve Vanderheiden
Professor of Political Science, University of Colorado Boulder

Shelley Welton
Presidential Distinguished Professor of Law and Energy Policy, The University of Pennsylvania

INTRODUCTION

CHIARA CORDELLI AND MELISSA LANE

In choosing climate change as the focus of the symposium that led to this volume, the members of the American Society for Political and Legal Philosophy recognized the threat that it poses not only to established modes of human existence but also to many established assumptions of the participating disciplines of politics, philosophy, and law. Convening in September 2023, soon after the publication of confirmation that "Over the 2013–2022 period, human-induced warming has been increasing at an unprecedented rate of over 0.2°C per decade,"[1] the symposium took stock of the ways in which each discipline has conceived the roots of climate change, and possible ways to address them, over the course of several decades. All three disciplines have variously grappled with issues of temporal and spatial scale and variation; with the location of moral, legal, and political responsibility; and with dramatic changes in the degree and distribution of risk and uncertainty, in thinking through the ways in which climate change is challenging established modes of social organization.

To be sure, there are distinct traditions of debate in each discipline. Much argument in philosophy has revolved around the question of responsibility to, and even the identities of, future generations. In political theory, issues of justice, both compensatory and distributive, have been at the fore, including the perspectives on environmental justice rooted in Indigenous communities. In law, issues of international law, novel legal mechanisms (such as for carbon trading), and planning law have been especially prominent.

Despite these differences in approach and emphasis, this volume demonstrates a striking convergence in the questions that have recently come to the fore in political theory, philosophy, and law alike. The volume focuses on three main questions surrounding the ethics and politics of climate change. The first, which we may call the *methodological question*, concerns the limits of welfare economics, and of its utilitarian logic, as arguably still the dominant policy approach to climate change, and one that has loomed correspondingly large in intellectual approaches to the problem in many disciplines as well. Having a precise account of the limitations of welfare economics is necessary to develop not only new and more promising policy approaches, but also a more accurate account of our responsibilities toward existing and future generations.

The second question, which we shall refer to as the *diagnostic question*, investigates the very nature of the problem of climate change. Is the latter better understood as a problem of economic inefficiency, distributive injustice, republican domination, or existential unfreedom? Can a diagnosis of climate injustice proceed independently of a critical analysis of the economic and power structures within which environmental harms are produced and reproduced?

Finally, there is the *responsibility question*, which can be unpacked in several sub-questions. Which agents or entities (e.g., individuals, collectives, corporations, or structures) should be held morally responsible for climate change? To whom is such responsibility owed, e.g., existing people or not-yet-existing future generations? How should such responsibility, however attributed and distributed, be discharged, e.g., through a reduction of individual consumption, collective efforts at technological innovation, or by imposing immigration restrictions to limit population growth in high-emitting countries?

All the contributions in this volume can be read as addressing one or more of the above questions. With regard to the methodological question, Douglas A. Kysar points out how welfare economics treats as fixed inputs into economic models issues that should instead be themselves subject to policy choice—e.g., whether and how quickly societies will decarbonize through technological progress, or which populations' policies will be adopted. Further, it relies on a "bullish view of natural resource abundance and sustainability," which sees damages to future generations as compensable through an undifferentiated accumulation of capital, thereby neglecting the question of ownership—who owns the resources that have been destroyed— even as it presupposes it. Most importantly, the welfare economics approach, Kysar argues, narrows our political imagination and biases it toward the status quo, by anchoring us to "an imagined world in which ownership and exchange under largely current conditions define what is achievable through politics." Hence the inability of most political actors to conceive of solutions beyond a carbon market. Against such narrowing, Kysar proposes two novel solutions to the problem of climate change—"carbon upsets" and inequality reduction. While not being incompatible with the dominant approach in economics, such solutions push the boundaries of our political imagination beyond the status quo.

Zeynep Pamuk, like Kysar, is critical of the fact that climate-focused economists and policymakers operate on the basis of scientific, and yet highly controversial, models that are shielded from proper public scrutiny and contestation. Her worries about this tendency toward depoliticization go even further than do Kysar's. The problem for Pamuk is not simply that such models constrain our political imagination but also that they do so in an undemocratic way. The broader question for Pamuk concerns the appropriate role of scientific expertise within a democratic polity. While Pamuk does not deny that experts exist or that they should play a significant role in climate policy, she argues for the necessity of institutions that, by enabling the public scrutiny and contestation of scientific evidence, would check and make more accountable the political power of climate scientists and economists—their power to shape and direct our political imagination and collective action. Like Kysar, Pamuk proposes novel solutions with an institutional dimension, including an adversarial "science court," where "proponents of green growth, degrowth, and a Green New Deal could debate one another in front of a jury of randomly selected citizens."

While also starting from a critique of (neoclassical) economics as guilty of reducing an existential threat such as climate change to "a problem of missing prices" or "externalities," Alyssa Battistoni's piece brings us to the diagnostic question, offering a reconceptualization of the very nature of the problem of climate change. Climate change, she argues, is neither a mere problem of exchange or market failure, nor simply an "aggregation problem," as often defined by philosophers, most prominently Derek Parfit; nor is it straightforward interpersonal harm, as has been emphasized by John Broome.[2] Rather, climate change is more accurately understood as a problem of structural domination in at least two respects, and so implicates not only justice, but also, and perhaps even more centrally, freedom.

First, within capitalism, where a class of people enjoys an institutionally sanctioned form of advantage over weaker classes, the (alienable, because tradable in so-called carbon markets) right to pollute becomes a means of domination: a way in which the dominant group is able to profit by arbitrarily imposing social costs on others, thereby reducing the latter's options for choice, and their freedom. The fact, for example, that formerly colonized peoples are more likely to accept trades in toxic waste and bear higher pollution burdens, Battistoni argues, is a function not of their free, voluntary choice, but of their structural vulnerability and market dependence. This in turn implies that "the distributive dimensions of climate change are not only matters of injustice but of unfreedom."

Second, markets make all of us unfree by making us irresponsible for our choices, including our environmental choices. Philosophical attempts to attribute moral responsibility for environmental matters to specific individual consumers often fail, according to Battistoni, for consumer choices are mediated by a market framework that fundamentally denies individual responsibility, by forcing all participants to make choices under conditions of ignorance, and by weakening the connection between intent and outcomes. Battistoni's piece ultimately speaks to the impossibility of addressing the problem of climate change independently of the capitalist organization of the economy, and of society at large.

Mark Budolfson rejects this conclusion. Contrary to Battistoni, he argues that the best solutions to climate change are perfectly consistent with capitalism, and actually require capitalism. Failures to address the problem of climate change should not be attributed to capitalism as such, but rather to the existence of certain forms of democratic politics, which fail to regulate capitalism appropriately. However, Budolfson agrees with Battistoni that standard economic policy often ignores, and even reproduces, problematic forms of structural injustice and inequalities. So, well-regulated capitalism requires more than standard economic policy.

Madison Condon, unlike Budolfson, is more sympathetic to Battistoni's view that addressing climate change requires moving beyond a capitalist organization of the economy. At the same time, she rejects Battistoni's tale of market deresponsibilization. Far from

Introduction 5

all being rendered irresponsible, and thus unfree, by the market, Condon argues that some actors, and in particular business corporations, have far more power, and thus also more responsibility for the climate crisis than others. The pernicious role corporations play in the climate crisis is, however, itself a consequence of how the law understands the corporation and its directors' duties, often and mistakenly regarding these duties as owed to shareholders rather than to the corporation itself. Condon argues for returning to a pre-neoliberal legal conception of the corporation as a moral entity, and as a site of political change in the fight against climate change. Corporate law, she points out, can and should enable investors to pressure corporations into decarbonizing. Indeed, she ingeniously suggests that, once we leave behind a commitment to shareholder primacy, even a narrow cost-benefit economic analysis could support a corporate imperative to decarbonize, as the long-term profit of the corporation, unlike the short-term profit of shareholders, is best served by prioritizing climate mitigation over fossil fuel extraction. In different ways, Condon's reflection on the privileged position of the corporation, and Battistoni's on the structural domination of capitalism, connect the diagnostic question (climate change as a problem of unequal sites of power) to the responsibility question—who should be held morally responsible and liable for climate change, and to whom such responsibility is owed, which is central to all remaining contributions.

Likewise focused on the question of responsibility, Lucas Stanczyk develops a novel way of understanding our obligations to future generations in highlighting the issue of intergenerational justice. His account is intended to be nonconsequentialist, as it focuses on respecting individuals' rights rather than on maximizing aggregate utility over time; and it is meant to be person-regarding, as it proposes interventions that would make particular individuals better or worse off than those same individuals would otherwise be without those interventions. According to Stanczyk's account, duties of intergenerational justice are owed by some of the adult members of the present generation to its younger members, who are already alive and who one day will become the adult members of a new generation, owing duties to its younger members, and so on (by eliminating duties toward not-yet-existing persons, Stanczyk aims to make Parfit's well-known Non-Identity Problem irrelevant). Further, the imperative of intergenerational justice for Stanczyk is not to maximize welfare but rather "to put in place all those restrictions on existing rights and freedoms that, unless they are put in place, will require even more serious sacrifices having to be made later by today's younger people, lest they fall afoul of the very same imperative in the future." Interestingly, then, the core of intergenerational *in*justice for Stanczyk is a form of "kicking the can down the road," by failing to restrict rights now (e.g., by severely limiting consumption or aviation) in ways necessary to prevent even more severe sacrifices having to be incurred later by the now young in order to secure every living person's basic interests. On a more practical level, Stanczyk defends the view that our obligations of intergenerational justice require policies directed at drastically reducing consumption and controlling population growth. Investing in greener forms of energy production is far from enough.

F. M. Kamm addresses the philosophical core of Stanczyk's theory of intergenerational justice by questioning whether his account is truly nonconsequentialist. Kamm points out that Stanczyk identifies which claims should be sacrificed for the sake of preventing harm to future generations on the basis of the strength of the interests such claims protect (e.g., severely limiting consumption of nonessential goods now in order to prevent imposing more severe sacrifices on the next generation). She argues, however, that this is incompatible with a deontological, nonconsequentialist approach. As she puts it, "it is not necessarily true that the one who should pay to prevent the harm is whoever would sacrifice a claim involving a lesser interest in order to prevent the harm." Kamm further doubts that Stanczyk can successfully overcome the Non-Identity Problem, by trying to limit duties of intergenerational justice to existing persons. All in all, Kamm's internal critique of Stanczyk calls attention to the requirements that an account of climate ethics should meet in order to qualify as genuinely deontological.

Steve Vanderheiden, like Stanczyk and Kamm, is also concerned with the question of responsibility, and with the problem of "can-kicking," but he endorses a perspective more akin to Kysar's on the primary role of the state (and international treaties and institutions) to exert control over particular aspects of the economy. He proposes that duties to mitigate climate change, and thus to reduce its harmful impact on future generations, should be articulated in terms of fair shares of an overall just carbon budget. The purpose of such a periodically revisable budget is to allocate remaining emittable carbon dioxide among groups and their members, as well as across generations, in consideration of broader demands of global justice, as well as changes in circumstances (e.g., technological development). In Vanderheiden's words, "a just allocation . . . would be one in which each agent did their part in a collective effort that satisfied imperatives of global as well as intergenerational justice." In his account, the bearers of primary duties are collectives, while individuals are mostly left with the responsibility to pressure collectives to do their parts. Vanderheiden, departing from Stanczyk, specifies the injustice of "kicking the can down the road" in terms of failing to act according to one's fair share of mitigation duties, rather than on a principle of harm avoidance.

Shelley Welton shifts the focus on the practical implications of Stanczyk's proposal. While she agrees that the climate crisis is unavertable without more attention to consumption and population, she argues for important qualifications on how such attention should be given. On the one hand, it is important to ask "*whose* consumption must change," as levels of consumption-based emissions tend to track background economic inequalities. Similarly, we should ask *whose* population growth impacts climate in a significant way. Adding one person in the United States is much worse, in emissions terms, than adding a person in sub-Saharan Africa, precisely because average consumption in sub-Saharan Africa is much less than average consumption in the US. A focus on controlling population growth in the Global South, beyond raising concerns of relational justice and domination well identified by many feminist and postcolonial theorists, may not even be an effective means of reducing emissions. On the other hand, Welton worries that invoking consumption and population-related concerns in climate discourse and climate policy may politically backfire. This observation raises the important question, "what role should climate ethics play in on-the-ground climate policy and politics?" Bringing together Samuel Moyn's concept of "situated freedom"—the space contained in the laws for "critique and transformation"—and Amna Akbar's account of "non-reformist" reforms, Welton suggests ways in which climate policy can and should be designed so as to open up possibilities for more radical dialogue and larger structural change.[3]

Also concerned with policy implications, Jamie Draper's contribution tackles a new aspect of the question of moral responsibility for climate mitigation: its interaction with the ethics and politics of immigration. Draper asks whether egalitarians concerned about climate change should support immigration restrictions on environmental grounds, since immigration to high-income countries can lead to an increase in polluting emissions, as it increases population in such countries. Draper argues that egalitarians need not support such restrictions. This is not because migration from low- to high-income countries does not have harmful environmental consequences (although Draper also argues that remittances may support efforts at climate adaptation in immigrants' countries of origin). It is rather because the consequences of immigration are harmful only because of actions taken by the receiving state (e.g., allowing high consumption standards). Draper argues that egalitarians should instead (for independent moral reasons) advocate for more open border policies (to be adopted for independent moral reasons) but that these must be coupled with a call for aggressive climate mitigation in high-income countries.

Beyond addressing the three fundamental questions of method, diagnosis, and responsibility, the chapters in this volume also contribute to offering novel policy solutions to the problem of climate change. While Kysar, as we saw, proposes both a scheme of "carbon upsets"—awarding credits to social movements and other groups that carry out efficacious forms of climate-preserving legal and political actions—and inequality reduction mechanisms (e.g., income and inheritance taxes) as desirable policy measures, Stanczyk points to the necessity of population control programs and of drastic caps on individual consumption. Finally, Draper suggests that policies aimed at liberalizing immigration can counterintuitively, in certain circumstances, function as an effective environmental strategy. In these ways and others, the volume pushes forward the complex and by now long-standing debate on climate ethics and politics.

Acknowledgments

This volume of *NOMOS* emerged from papers and commentaries given at the annual meeting of the American Society for Political and Legal Philosophy, held at the University Center for Human

Values at Princeton University in September 2023. The topic of the volume, "Climate Change," was selected by the Society's membership. This volume includes revised versions of the principal papers delivered at the conference by Douglas A. Kysar, Alyssa Battistoni, and Lucas Stanczyk, as well as essays that developed from the commentaries on those papers by Mark Budolfson, Madison Condon, Zeynep Pamuk, Steve Vanderheiden, and Shelley Welton. The volume also includes solicited essays by Jamie Draper and F. M. Kamm. We are grateful to all the authors for their excellent contributions as well as to all participants in the conference discussions and the University Center's staff for their support. Thanks also to Cole Smith for his meticulous and thoughtful editorial assistance.

We are also grateful to the editors and production team at New York University Press, particularly Alexia Traganas and Sonia Tsuruoka.

Finally, we thank the members of the ASPLP Council: President Deborah Hellman and Immediate Past President David Estlund, Vice-Presidents Anita Allen and Anna Stilz, Secretary-Treasurer Micah Schwartzman, Outgoing Editor Eric Beerbohm, Communications Director Jennie Ikuta, and At-Large Council Members Brandon Terry and Leif Wenar. As ASPLP's current Editor, Chiara Cordelli shouldered the lion's share of the work on this volume, and Melissa Lane is grateful to her for that and for the opportunity to join together in this project.

Notes

1 Piers M. Forster, Christopher J. Smith, Tristram Walsh, William F. Lamb, Robin Lamboll, Mathias Hauser, Aurélien Ribes, et al., "Indicators of Global Climate Change 2022: Annual Update of Large-Scale Indicators of the State of the Climate System and Human Influence," *Earth System Science Data* 15, no. 6 (2023): 2295–2327, at 2296.

2 Derek Parfit, *Reasons and Persons* (Oxford: Oxford University Press, 1984); John Broome, "Against Denialism," *The Monist* 102, no. 1 (2019): 110–129.

3 Samuel Moyn, "From Situated Freedom to Plausible Worlds," in *Contingency in International Law: On the Possibility of Different Legal Histories*, ed. Ingo Venzke and Kevin Jon Heller (Oxford: Oxford University Press, 2021). Amna A. Akbar, "Non-Reformist Reforms and Struggles over Life, Death, and Democracy," *Yale Law Journal* 132, no. 8 (2023): 2360–2657.

PART I
THE LIMITS OF CLIMATE ECONOMICS

1

WAYS NOT TO THINK ABOUT CLIMATE CHANGE

DOUGLAS A. KYSAR

A half-century ago, famed constitutional law scholar Laurence Tribe wrote an article discussing humanity's relationship to the natural environment. It was entitled "Ways Not To Think About Plastic Trees: New Foundations for Environmental Law."[1] The article was then and remains today a brilliant discussion of the root intellectual causes of contemporary environmental crises. Indeed, it is no overstatement to say that environmental law scholarship in the United States over the subsequent five decades consists largely of refinements and applications of the insights and debates that Tribe framed in the article's thirty-three pages. At the outset, Tribe identified a powerful but often unexamined set of assumptions that seemed ascendant concerning the appropriate way to design and pursue environmental policy. As he wrote, "[t]hese assumptions, which are implicit in developing uses of policy analysis as well as in emerging institutional structures, make all environmental judgment turn on calculations of how well human wants, discounted over time, are satisfied."[2]

As part of his agenda in the article, Tribe articulated several conceptual and methodological limitations of this narrow welfare economic mode of evaluating humanity's environmental role—limitations that remain largely unanswered today despite the enormous amount of academic and institutional attention devoted to regulatory cost-benefit analysis over the ensuing time period. As powerful as they were, these critiques were not the most significant contribution of "Ways Not To Think About Plastic Trees."

Instead, the article's most prescient and weighty warning concerns the potentially destructive *endogenous* effects of an environmental policy framework that is oriented narrowly around human welfare and preference satisfaction, irrespective of whether the internal shortcomings of the framework might be alleviated:

> What the environmentalist may not perceive is that, by couching his claim in terms of human self-interest—by articulating environmental goals wholly in terms of human needs and preferences—he may be helping to legitimate a system of discourse which so structures human thought and feeling as to erode, over the long run, the very sense of obligation which provided the initial impetus for his own protective efforts.[3]

The golden anniversary of Tribe's seminal contribution to environmental thought is an important moment to take stock of its influence.[4] In brief, rather than acknowledge the inevitable coevolution of means and ends that Tribe identified—which forces policymakers to engage at least somewhat with the question of what kind of culture and values they wish to promote through their decisions—the United States seems instead to have moved ever further in the direction of a cabined, anthropocentric, and rigidly economic vision of environmental policymaking, so much so that leading scholars have referred approvingly to the country as "The Cost-Benefit State."[5] This trend has held largely true across nine presidential administrations, from Ford to Biden, irrespective of party affiliation and despite the dramatic polarization of environmental issues along party lines that opened over that time.

It also has held true despite the rise of climate change as an environmental challenge unlike any other created by Earth's currently most dominant species. Recognized by the US national government as a profound, even existential threat as early as 1965,[6] climate change nonetheless has tended to be analyzed, like any other environmental ill, through the lens of welfare economics.[7] Proponents argue that cost-benefit analysis of climate policy is necessary to ensure that societies do not overinvest in environmental protection, as if "excessive" environmental precaution were a genuine societal risk within the Great Acceleration—the current period of unprecedented rise in human population and corresponding surge in numerous socioeconomic and ecological measures of human activity and its impacts. Supporters of climate economics even sometimes insist—in categorical terms, based on the asserted need to prioritize efficiency when reducing emissions—on market-based instruments such as tradeable permits and carbon pricing for climate policy, as if transitioning from an unsustainable pathway *at least cost* is more important than transitioning *well before* passing nonlinear thresholds that carry potentially cataclysmic consequences. To critics, these efficiency rationales obscure the deeper way in which neoliberal climate thinking is insensitive to historical and moral context.[8] Some further worry that neoliberal climate thinking risks entrenching the very systems that have brought the world to the brink of disaster.[9] Indeed, as discussed below, influential climate economic models cheerfully contemplate within their framework the possibility of greenhouse gas emissions pathways that entail the near-certain extinction of the human species, yet somehow preserve the bulk of global economic production long after humans may be gone.

Guided by such a way of thinking, it is perhaps not surprising that government leaders in the United States have achieved relatively little in terms of domestic climate change legislation and regulation, leading learned observers to argue the framework has exhausted whatever utility it may once have had.[10] Internationally, matters are similarly bleak: In the three decades since the 1992 international climate convention was signed, humanity has released more carbon into the atmosphere than during the rest of human history combined. In that respect, it is at least encouraging that the most recent four years of US climate policy have signaled something of a shift away from obsession with neoliberal climate thinking in favor of unabashed industrial policy, transitional and environmental justice, and familiar, road-tested regulatory instruments, as evidenced by passage of the Inflation Reduction Act, adoption of the Justice40 Initiative, and efforts to implement the Clean Air Act and other existing statutory sources of agency authority in the face of an increasingly obstructionist Supreme Court. Much of the credit for that shift of narrative can be attributed to environmental justice, climate youth, and allied social movements that have refused to engage the climate change issue only in the narrow terms dictated by neoliberal climate economics.

Still, despite the inadequacy of regulatory responses flowing from the welfare economic framework and despite hopeful signs that the policy community may be ready to tip into a more expansive mode of engagement with the climate crisis, Tribe's fundamental challenge regarding the coevolution of means and ends in environmental law remains a pressing theoretical and practical question. During these decades of what might be called neoliberal hegemony in climate policy, what has been lost and what has been gained *within us*, putting aside what government actions may or may not have been pursued and what climate progress may or may not have been achieved? Consider another instance of the discerning warning regarding such endogenous cultural implications that Tribe offered back in 1974 and that merits earnest reflection from the vantage point of today, especially in relation to climate change:

> [M]ost of the crucial environmental choices confronting industrialized nations in the last third of the 20th century will be choices that significantly shape and do not merely implement those nations' values with respect to nature and wilderness. Such choices will do more than generate a distribution of pay-offs and penalties to the persons affected in terms of their preexisting yardsticks of cost and benefit. Choices of this type will also greatly alter the experiences available to the affected persons, the concomitant development of their preferences, attitudes, and cost-benefit conceptions over time, and hence their character as a society of persons interacting with one another and with the natural order.[11]

Over the last fifty years, has our "character as a society" been impacted by the dominance of neoliberal economic thought, again putting aside whatever influence it has or has not had on policies themselves?

This chapter examines specific features of the conventional welfare economic approach to climate change—primarily the social cost of carbon and carbon offset mechanisms—with a view toward identifying ways in which those features shape and constrain imagination. It bears emphasizing that within the economics profession, there have long been those who questioned the wisdom and feasibility of climate optimization models. It also bears noting that climate economics is now a vast and diverse literature with many models and approaches that avoid some of the methodological pitfalls explored below. Finally, those pitfalls themselves often can be traced to intellectual features of larger trends—such as the individualism, presentism, and anthropocentricism of the liberal tradition within which economics sits—that perhaps make for more impactful targets of criticism.[12] Despite those caveats, this chapter argues that the dominant climate economic framework in key respects leaves us morally and politically blinkered, unable to perceive and address some of the most significant issues raised by the climate crisis. As a result, it is at least arguable, as Tribe surmised, that our ability to ethically engage with the climate crisis has been stunted by the very tools we have brought to the task. Our effort to promote "instrumental rationality" has carried with it an implicit worldview, one in which our "ends are exogenous, and the exclusive office of thought in the world is to ensure their maximum realization, with nature [merely] as raw material to be shaped to individual human purposes."[13]

In addition to identifying the ethical narrowing wrought by conventional welfare economics, this chapter also lays out two policy proposals that, despite being readily implementable within the conventional framework, nevertheless contain within them seeds of dramatic shifts in thought. The fact that the proposals have not been discussed widely to date suggests the limiting power of our conceptual anchoring in neoliberal climate thought. The hope, though, is that situating the two proposals within that familiar and all-encompassing framework might help to denaturalize it, revealing some of the arbitrary constraints that are built into it and perhaps allowing fundamental aspects of climate policy to be confronted more directly and with less confusion.

Climbing the Wrong Mountain

At the core of environmental issues are questions regarding our moral obligations to people residing in other nations, members of future generations, and nonhuman life, all of which are deeply affected by choices and actions that are largely outside those groups' influence or control.[14] In most countries, one must add domestic victims of environmental injustice—presently living but politically oppressed people who suffer disproportionately from the negative effects of economic activity. The fundamental and unavoidable moral questions raised by these relationships must be addressed *before* turning to objective decision-making mechanisms such as economic cost-benefit analysis or their related implementation devices. Yet those mechanisms and devices often give the appearance of having somehow already addressed and resolved the underlying moral questions, leaving observers falsely reassured that all relevant aspects of environmental decisions can be properly analyzed within the decision-making framework. Vital moral aspects of environmental law and policy become obscured and the ability of ordinary people to appreciate and participate in the process of future-making becomes occluded. All the while, the planet burns.

The recent history of national environmental law and policy in the United States illustrates these dangerous dynamics. Two presidents—Barack Obama and Donald Trump (referring here to his first administration)—took diametrically opposed positions and actions on climate change and other important environmental issues, yet both justified their positions and actions by appealing to economic cost-benefit analysis. In practice, the "game" of regulatory cost-benefit analysis in the United States has become just that, a structured exercise in which competing interests pursue policy outcomes not through direct argument and suasion, but through use of alternative modeling assumptions, valuation techniques, discount rates, and other seemingly technical trappings of the cost-benefit methodology. As a result, subjects of ordinary moral and political discourse are debated through a stylized cost-benefit vernacular that both enables power and renders it illegible.[15]

This criticism holds even for a nominally pro-environment president such as Barack Obama, who initially campaigned on a platform that included strong environmental messages and a commitment to engage the climate change problem promptly and aggressively. In the opening moments of his administration, President Obama also signaled a strong desire to "mend it, not end it" when it comes to economic cost-benefit analysis,[16] both by nominating noted cost-benefit proponent Cass Sunstein to serve as the country's "regulatory czar" and by issuing an announcement that regulatory cost-benefit analysis would be retained in his administration, subject to some modest changes to acknowledge progressive goals such as incorporation of distributional and fairness concerns, respect for the interests of future generations, and avoidance of undue delay in rulemaking.[17] Unfortunately, the global economic recession of 2009 also coincided with these developments and worked to destroy all apparent political appetite in the United States for environmental, health, and safety reform, including especially with respect to climate change. President Obama instead prioritized health care for his first-term legislative agenda and was forced to resort to executive actions only to address climate change during his second term.

Those executive actions were guided significantly by the "holy grail of climate economic analysis"[18]—an analytical device known as the social cost of carbon which purports to capture in monetary terms the negative impacts of each additional ton of carbon dioxide (CO_2) or equivalent greenhouse gas emissions. The social cost of carbon in theory enables regulators to justify mandating emissions reductions by showing that the cost of doing so for industry is less than the harms that would be imposed on society without reductions. To develop a uniform federal measure of the social cost of carbon, an interagency task force under President Obama consulted existing integrated assessment models from the academic literature, such as Nobel laureate William Nordhaus's influential DICE model.[19] In its resulting report, the task force acknowledged serious limitations and shortcomings in the cost-benefit methodology as applied to the climate change problem. Some of these moments of candor and humility even broached fundamental ethical subjects—such as whether it is appropriate at all to discount the interests of future generations—that many cost-benefit proponents have tended to avoid. For these reasons, the task force should be applauded.[20]

Nevertheless, after contemplating its various limitations and shortcomings, it is hard not to be left wondering whether a social cost of carbon estimate is useful at all.[21] To be sure, US federal agencies are under an executive branch mandate to conduct cost-benefit analysis of major proposed rules, and courts have held that such analyses must include a quantitative estimate of the benefits of avoiding further climate change when applicable and feasible.[22] As a practical matter, analysts also frequently contend that having some number, however imperfect, is better than having no number at all. But is it? After all, "[a] bad analysis can be so wrong that it can lead us to do bad things, outrageous things—things that are much worse than what we would have done had we not tried to assess the costs and benefits at all."[23] For that reason, one might conclude that the social cost of carbon is simply the wrong tool for the climate change job. Indeed, the social cost of carbon is a tool that contains—buried deep within its assumptions—deceptively narrow and limited answers to the most fundamental moral and political questions raised by climate change. As explained below, this was the case even before the first Trump administration took office and turned economic cost-benefit analysis into an unabashed exercise in power and manipulation, rather than anything remotely displaying academic rigor and objectivity.

To understand the core shortcoming of cost-benefit analysis in the environmental context, imagine the pursuit of social-welfare maximization as a quest to climb a mountain. By evaluating proposed changes to the status quo in terms of incremental welfare consequences, cost-benefit analysis promises to determine whether any given policy change will lead marginally higher up or down the mountain of social welfare. However, the fundamental problem in the climate change context is that cost-benefit analysis cannot tell us *whether we are on the right mountain*. While scrambling meticulously over the details of any cost-benefit exercise, it is easy to lose sight of the fact that at its base lies a set of fundamental assumptions about resource rights, income distribution, population size, intergenerational equity, international obligation, the likelihood of technological innovation, the capacity of individuals and communities to adapt to climate change, the trajectory of the economy, and so on. Alter these assumptions and one stands on a different mountain, where cost-benefit analysis once again can offer advice on whether a proposed step will lead incrementally up or down.

In the climate change context, the near universal view of natural scientists is that we are currently on the wrong mountain. Indeed, because of tipping points in a variety of planetary systems, we are heading up a mountain with a cliff at its peak.[24] Thus, we are well advised at present to take policy steps that may appear *inefficient* when measured by marginalist cost-benefit analysis that is fixed to the assumptions of the status quo. Shuttering existing coal-fired power plants and capping still-productive oil wells are examples of actions that would likely fail a conventional cost-benefit analysis today but that might well be the most rational actions humanity could take from a perspective that sees beyond our current mountain. In the limited terms of integrated assessment models like DICE, we may have to climb *down* for a while before we can again start ascending a different peak. The ultimate result will be a mountain with a better and more enduring view, but the path of transition will not be one dictated by marginalist cost-benefit analysis.

The Obama task force report did stress this fundamental distinction between choosing and climbing a mountain, or to put the point more technically, between comprehensive and marginal analysis. Nevertheless, the authors confidently stated in the report that their marginal social cost of carbon estimate will be useful because "[m]ost federal regulatory actions can be expected to have [only] marginal impacts on global emissions."[25] The danger here is that the authors' assertion will become self-fulfilling: Whether an environmental law offers marginal or infra-marginal possibilities for altering greenhouse gas emissions trajectories depends very much on whether the environmental regulator decides to implement the statute with the vigorous, transformational teeth it was designed to have. For instance, when viewed individually, energy investment decisions such as whether to license new fossil fuel-fired power plants may appear to represent only marginal alterations to a status quo trajectory of emissions growth. But when one considers the durability of such investments combined with the fact that they occur within a national policy space in which dozens of such individual decisions are constantly being made, then the licensing decision may well be seen to raise an *infra*-marginal policy concerning the future trajectory of a nation's energy infrastructure.[26]

The integrated assessment models relied on by the task force are built on assumptions about matters that are central to the climate change problem, but that are not allowed to surface for direct inspection. The cost and benefit outputs of the models depend critically on assumptions about such matters as whether and how quickly economies will de-carbonize through technology innovation, what level of adaptive capacity communities will hold to lessen the impacts of climate change, what population policies nations will pursue, what level of international action can be anticipated to follow from domestic leadership, and so on. These matters are treated as fixed inputs into the models rather than as *themselves* subjects of policy choice. As a result, regulations altering the trajectory of these matters may be assessed using the social cost of carbon, but the apparent "costs" and "benefits" of such regulations will be misleadingly calculated. To give just one example, none of the models relied on by the task force allow for the possibility that major shifts in US policy will cause other nations to follow suit, such that the global emissions trajectory will be lowered by an amount greater than the direct domestic impacts of the US policy change. Accordingly, the apparent benefits of US climate policy proposals will be systematically understated by the models if it is true—as countless knowledgeable observers suggest—that US recalcitrance has long been a major stumbling block to international climate progress.[27]

Likewise, the foundational ethical questions at the heart of cost-benefit analysis—such as the debate over how to value lives or the best way to consider the rights of future generations—are typically treated by cost-benefit proponents as matters of elite expertise or disciplinary orthodoxy, rather than debatable moral and political issues. As a result, the prevailing principles and assumptions of cost-benefit analysis are rarely subjected to transparent and sustained critique, even though they often work to stack the deck against aggressive regulatory action on behalf of environmental, health, and safety protection. Even the Obama task force, which in many respects was a model of open and self-critical disciplinary examination, nevertheless balked at key moments in its analysis just as foundational questions were coming into view. For instance, after noting that the practice of discounting future costs and benefits to a present value raises fundamental ethical questions regarding the care and consideration present generations owe to future generations,[28] the task force authors nevertheless concluded that discounting should be applied when calculating the social cost of carbon simply to maintain "consistency with the standard contemporary theoretical foundations of benefit-cost analysis."[29]

The particulars of the discounting debate have been reviewed at length elsewhere and will not be recounted here.[30] For present purposes, it is enough to note that discounting is a crude and misleading way to incorporate matters of intergenerational ethics into the welfare-maximization exercise. If climate change policies must be evaluated using discounted cost-benefit analysis, then analysts should do so using shadow markets in which natural resources and other environmental goods are *first* endowed to future generations through sustainability constraints or other hypothetical regulatory measures.[31] Modeling a sustainable market economy in this way will generate an interest rate that regulatory analysts might *then* use to adjust future costs and benefits when evaluating policies. Significant intergenerational policy issues like climate change will look dramatically different when evaluated in this way. If we first endow future generations with the right to a relatively safe and stable concentration of greenhouse gases in the atmosphere—as physical scientists implicitly urge when they propose policy targets such as a maximum 1.5°C temperature increase over pre-Industrial levels or a 350 parts per million CO_2 concentration stabilization—then the present generation will be required to "purchase" that endowment from future generations using prices that reflect a normatively defensible background allocation of atmospheric rights. As it is, conventional cost-benefit analysis proceeds as if the present generation owns everything, implicitly answering without analysis a question that lies at the very core of the climate change policy problem. Importantly, this question cannot be avoided even if one adopts the mainstream welfare economic view that future generations can be "compensated" through undifferentiated accumulation of capital whenever specific natural resources are compromised—even a resource as basic and seemingly non-substitutable as atmospheric stability. Even on this bullish view of natural resource abundance and substitutability, the level of "compensation" that is due to future generations must be calculated in a way that *first* asks who owns the resources that have been destroyed. Any other procedure allows the present generation to play lord in a game of temporal feudalism.

Analytical confusion in the realm of climate economics carries grave consequences. Even the Nobel Prize–winning DICE model can be criticized for overstating the bounds of its competency. As figure 1.1, taken from the Obama task force report illustrates, within the DICE model—like the other two models relied upon by the task force—global temperatures can increase to unfathomable levels without fundamentally affecting world economic output. According to DICE, only one-quarter of global GDP would be lost if temperatures increased 10°C above pre-Industrial levels. Indeed, only one-half of global GDP would be lost in the DICE model at a temperature increase of 19°C.[32] Consider this outcome in light of scientific calculations of a climate change "adaptability

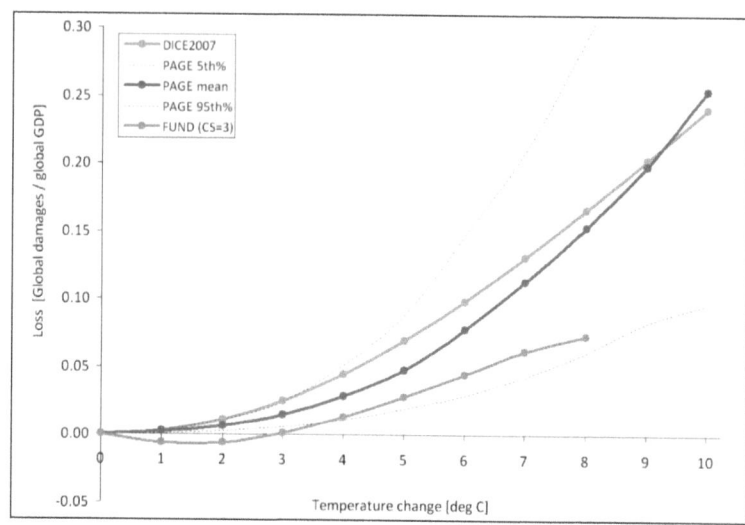

FIGURE 1.1. Annual Consumption Loss as a Fraction of Global GDP in 2100 Due to an Increase in Annual Global Temperature in the DICE, FUND, and PAGE Models. Source: Interagency Working Group on Social Cost of Carbon, US Government, "Social Cost of Carbon for Regulatory Impact Analysis Under Executive Order 12866," Technical Support Document, Washington, DC.

limit" due to the human body's inability to dissipate heat and avoid hyperthermia above a certain temperature. Scientists studying this limit "conclude that a global-mean warming of roughly 7°C would create small zones where metabolic heat dissipation would for the first time become impossible, calling into question their suitability for human habitation. A warming of 11–12°C would expand these zones to encompass most of today's human population."[33]

How can it be that one-half of global GDP would continue to be generated and enjoyed at a temperature level several degrees Celsius above an adaptability limit for human survival? Quite simply because harm of that sort is an infra-marginal event not contemplated by the model's damage function. Rather than positing some degree of fundamental dependence between socioeconomic and natural systems, integrated assessment models typically assume that the economy will continue to function more or less as is, even as damages from climate change grow ever larger.[34] In the extreme, this means that global GDP can continue to pour forth within the models even after all presently inhabited land on Earth has been rendered unsuitable for human existence. The possibility of such extreme climate change scenarios suggests that cost-benefit optimizing may simply be the wrong framework for analyzing climate change policy. Rather than optimal consumption smoothing over time, policymakers instead should consider climate change from an insurance perspective, asking how much of present consumption is worth investing in the avoidance of truly intolerable outcomes.[35] Or, to give the question a more explicit moral valence, policymakers should ask whether the present generation wants to countenance global disaster among its legacy of achievements. Because catastrophic climate change scenarios are essentially uninsurable and intolerable events, this mode of thinking leads to a goal of prevention along lines the scientific community has been urging all along: Rather than a system in which all resources are optimally deployed to their highest-valued use according to some set of fundamentally contestable assumptions, we should seek a system that displays characteristics like precaution, diversification, resilience, redundancy, and innovative capacity—characteristics that may appear to represent an "inefficient" use of resources from a narrow economic viewpoint, but that may be essential nonetheless.

Recall again that these integrated assessment models, with all their flaws and confusions, were embraced by Obama, a nominally *pro*-environment president. One of the first executive actions undertaken by the Trump administration (elected in 2016) was to disband the interagency task force and disavow the social cost of carbon report. Administration officials quickly made clear that the "game" of cost-benefit analysis would be played differently under Trump, lowering the social cost of carbon from the Obama administration's mean estimate of US$50 per ton to US$1–7 per ton.[36] Three changes in particular worked to achieve this lowering and, in turn, to dramatically constrict the government's ability to address climate change and other environmental ills. First, the Trump administration adopted a much higher social discount rate in order to lower the apparent value of undertaking climate mitigation policies. Second, the administration barred agencies from considering any benefits that happen outside US borders when considering the costs and benefits of environmental regulations. Third, agencies also were prohibited from weighing "co-benefits," i.e., supplementary benefits that occur in addition to the main target of a regulation, such as mercury emission reductions that occur when CO_2 limits are mandated for the utility sector. With these and other methodological adjustments, the Trump administration was able to pursue the unraveling of nearly every federal action on climate change and environmental protection undertaken by President Obama, while still justifying the attempted rollbacks as social-welfare maximizing according to cost-benefit analysis.[37]

The pendulum of cynicism swung back toward earnest academic and technocratic engagement under President Biden. Informed by a blue-ribbon academic report,[38] and by evolving thinking within significant quarters of the environmental economics community,[39] the Biden administration undertook a substantial review of the social cost of carbon. The Obama-era estimates of approximately $51 per ton adjusted for inflation were reinstated for immediate agency use, pending guidance from a new interagency task force that was constituted to update the work of the Obama social cost of carbon experts alongside a broader effort by the Office of Management and Budget to update executive branch oversight of regulatory cost-benefit analysis across all agencies and tasks. Advances in empirical estimation of climate damages and more refined understanding of how climate change poses disproportionate risks to poor and marginalized communities were just two of the improvements expected to emerge from this process.[40] Somewhat surprisingly, the Environmental Protection Agency under Biden proposed an estimate of about $190 per ton for a social cost of carbon to the interagency working group, but also made that figure and supporting analysis publicly available as part of a Clean Air Act rulemaking process for methane emissions from the oil and gas sector.[41] Observers speculated that the agency hoped to put political pressure on the task force to ensure that it would incorporate the latest scientific and economic research, which justified a much higher social cost of carbon estimate than even the Obama-era figure.

For some readers, these updates might feel like progress. But the progress remains contingent since the underlying conceptual limitations of the social cost of carbon have not been resolved. Consider the case of extraterritorial effects. Unlike the first Trump administration, which saw an opportunity to curtail federal agency regulatory "budgets" by simply excluding such effects, the Obama task force instead calculated both a domestic and a global social cost of carbon without choosing between them.[42] The EPA under President Biden chose to include global lives lost within a single revised social cost of carbon, but only monetized the value of foreign lives according to national income measures, rather than following the agency's standard practice of using a uniform monetary value of a statistical life. National income measures value the lives of most of the world's population at a significantly lower dollar amount than US lives. As legal scholar Daniel Hemel observed, applying a uniform value of a statistical life globally would have resulted in much higher estimated benefits from climate regulation, a level the agency presumably was unwilling to venture as a political proposition given that it had already come in supposedly "high" at $190 per ton.[43] These shifting approaches have a certain kind of half-transparency about them, but they share a common defect: In each case, the profound moral question of whether the United States owes consideration to foreign countries in respect of its contributions to climate change is treated merely as a modeling assumption to be decided by economic experts, rather than as a subject of utmost importance for political discussion and resolution.

At least in the United States, regulatory cost-benefit analysis often feels like lobbying in a different, more specialized vernacular—politics by other meanings.[44] It is a language spoken by few and dominated by even fewer. Its diction is poor though it purports to speak for everything. The fact that even Donald Trump's destructive environmental policies could be "justified" through cost-benefit analysis suggests that the enterprise has failed and that a return to fundamentals—to the openly moral and political questions that are raised by our relationship to the planet and its living beings—is necessary. Literally everything is at stake.

Counterfactual Carbon

As noted in the previous section, climate change policy is inseparable from foundational issues concerning international responsibility and coordination. For more than three decades, the deadlock on climate change policy at the global level has reflected a fundamental divide between developed and developing countries over how to allocate responsibility for greenhouse gas emissions reductions. Generally speaking, poorer countries—even major emitters such as India, Indonesia, and Brazil whose absolute emission levels are high but whose per capita levels remain lower than most developed countries—want to avoid aggressive reduction targets until they have achieved a level of development closer to that of wealthy nations. Many developed countries, for their part, are reluctant to take on stiff reduction targets because they fear a sucker's payoff: Whatever reductions they achieve may result in little climate benefit if highly populous and rapidly industrializing countries continue their massive growth in emissions.

One way that policymakers have attempted to bridge this divide is through use of carbon offsets. Under this approach, power plants, factories, and other regulated emitters in developed countries can satisfy their reduction requirements by purchasing carbon credits awarded through some official process, such as the Clean Development Mechanism under the Kyoto Protocol. Credits are awarded when a project undertaken in a developing country is less greenhouse gas–intensive than it would have been in the usual course. For instance, if a coal-fired power plant would have been built but financial support from a project sponsor instead leads to a lower-emitting natural gas plant, then the project can earn credits representing the difference in emissions levels between the respective technologies. Those credits can then be sold in the international carbon market for use by entities subject to emissions limitations.

The idea behind the offset approach is that incrementally shifting development onto a cleaner path in poorer countries is likely to be a much cheaper way to reduce emissions than direct cuts in richer countries. In theory, the approach also has the benefit of channeling additional development dollars to poorer nations. In practice, however, critics point to numerous problems with offsets. For instance, documenting what would have happened in the absence of financial support is far more of an art than a science, though the acceptability of the offset system depends on its appearance as the latter: "The calculational imperatives of the market ... dictate that the counterfactual without-project scenario be presented not as indeterminate and dependent on political choice ...

but as singular, determinate and a matter for economic and technical prediction."[45] Accordingly, a complex field of "private sector science" has developed to lend apparent technical legitimacy to the international offset system.[46] Project consultants within this field engage in a practice of "counterfactual display," in which "two future states of the world—one with the project and one without it—are played against each other and ... the value of the project is derived from that interplay."[47]

As with any such "centre of calculation," the political implications are both deep and obscure.[48] Numerous offset credits, for instance, are said to represent "anyway" credits, since projects receiving them likely would have happened regardless of outside financial support.[49] The result is a sheer wealth transfer without any actual climate change benefit and often with serious local environmental and social costs for the communities visited with the sponsored projects. Worse than "anyway" credits are the substantial offset rewards that have been given to firms that committed to destroying hydrofluorocarbon-23 (HFC-23), a potent greenhouse gas that results when chemical companies produce a refrigerant and propellant called hydrochlorofluorocarbon-22 (HCFC-22). Researchers determined that, in many cases, firms were building new factories or otherwise raising HCFC-22 production levels solely in order to create more HFC-23, which they then reduced in order to claim offset credits.[50] Incentives work ... even perverse ones.

Additional practical problems with offsets abound. Looming over all nature-based offset projects such as forest preservation and wetlands restoration are concerns about permanence, especially as climate-enhanced wildfires, hurricanes, and other threats render the character and stability of natural landscapes increasingly vulnerable. But more prosaic threats to offset integrity also exist. For instance, in the case of forest preservation and management, project sponsors can overstate existing deforestation trends, understate the risk of leakage (whereby prohibited emissions activities simply relocate elsewhere), inflate project sequestration capacity, undertake inadequate monitoring, and adopt other strategic practices that cast considerable doubt on whether any "additional" carbon actually has been stored by virtue of a project's registration.[51] An analysis of academic research concerning rainforest carbon offsets registered by a leading certifier determined that 94% of purported carbon reductions did not occur, in large part because background deforestation rates had been overstated by as much as 400%. In other words, the counterfactual display at the heart of the offset methodology opened opportunities for manipulation. As Oxford ecologist Yadvinder Singh Malhi put it, "[t]he challenge isn't around measuring carbon stocks; it's about reliably forecasting the future, what would have happened in the absence of [the offset incentive]. And peering into the future is a dark and messy art in a world of complex societies, politics and economics."[52]

Because of examples like this, carbon offsets have struggled under a cloud of suspicion since their invention. Today, even mainstream economists and governments have grown wary of carbon offsets because of the practical difficulty of ensuring their integrity. This is especially the case with offsets generated from projects abroad where compliance monitoring depends on third party verifiers who labor under conflicts of interest generated by their fundamental desire to see the carbon offset market thrive. As a result, the European Union has chosen to phase out reliance on international carbon credits within its emissions trading scheme, and other major carbon trading jurisdictions like California sharply limit their use. With greater attention to environmental justice, policymakers also have come to appreciate that outsourcing emissions reductions via offset markets often results in a lost opportunity to address serious domestic environmental problems that are often inequitably distributed along race and poverty lines. Greenhouse gases often are emitted from facilities that cause a variety of other, more locally harmful environmental ills. Allowing such facilities to satisfy their greenhouse gas emissions reduction responsibilities via carbon offsets leaves co-pollutants and other harmful impacts unaddressed.

Still, even if concerns about co-pollutants and environmental justice were addressed, and even if the various practical problems with carbon offsets were overcome, a more fundamental critique of the carbon offset approach would remain: The approach fails to incentivize the kind of dramatic, structural transformation toward a low-carbon future that is needed. "Carbon markets . . . serve as creative new modes of accumulation, but are unlikely to transform capitalist dynamics in ways that might foster a more sustainable global economy."[53] Even offset defenders acknowledge that the approach at best can achieve only minor changes to business-as-usual development. Yet the world's problem today is that business-as-usual development—even a marginally improved version of it—is a fast train to disaster. The impotence of the offset system in this respect is driven by its dependence on a narrow, essentially neoliberal imagination. Carbon offsets are a compliance device that proponents argue is necessary for the establishment of an efficient global carbon market, which proponents further argue is the only practical way to address climate change. These presumptions work to reinforce one another, as the actors operating within the carbon market become more and more anchored to an imagined world in which ownership and exchange under largely current conditions define what is achievable through politics. This narrow political imagination works to the exclusion of alternatives in which collective decision-making shapes the preferences, values, equity conditions, technologies, and structures within which we subsequently conduct market exchanges.

Carbon Upsets

What is especially fascinating about carbon offsets is that they explicitly rely on imagination; they are, in essence, "counterfactual carbon,"[54] legal instruments designed to represent and monetize the emissions that would have existed in a hypothetical business-as-usual-world without the intervention of an agent who is credited with having shifted downward the collective carbon trajectory. Once one admits the possibility of counterfactual carbon as a basis for distributing economic rewards and behavioral incentives, there should be no limit to the kinds of mitigation schemes one could concoct. Yet the offset system remains tightly anchored to a business-as-usual development vision: "While understanding what 'could have happened' in the absence of each particular project does mean funneling intellectual effort into speculation about hypothetical worlds, the hypothetical worlds that are relevant to determining 'what would have happened' without any particular project will all necessarily closely resemble the world with the project."[55] In this respect, the offset system seems politically palatable precisely because it is so consonant with the standard neoliberal path of finance and development: "[T]he logic of capitalization determines to a great extent the template of the imagined possible worlds."[56]

What would an offset scheme look like that could reside within a neoliberal framework without being limited by its imagination? Consider a system of what we might call "carbon upsets."[57] Rather than award credits based on economic development that moves us from an imagined dirty path toward a marginally cleaner but still very dirty future, why not award credits to legal and political actions that force downward our emissions pathway? Lawsuits, referenda, protests, boycotts, civil disobedience—these interventions too can lower emissions trajectories just as financial investments in offset projects supposedly do. For instance, a 2024 judgment of the Montana Supreme Court in favor of youth climate plaintiffs represented by Our Children's Trust will have a positive quantifiable impact via millions of tons of carbon equivalent emissions that the state could have authorized had the case not been brought. Likewise, a coalition of environmental justice groups in Louisiana is seeking to block approval of a massive petrochemical plant that would emit 13.6 million tons of carbon per year. If these legal actions succeed, they will not result in carbon credits despite their very real achievements. Under the neoliberal offset approach, "[p]olicymakers, environmental movements, indigenous communities who have prevented oil extraction in their territories: all have arguably saved carbon, yet are excluded from selling credits."[58]

Under a carbon upset system, credits could be awarded directly to groups and individuals when they work to achieve climate progress on their own. In addition, as with the existing offset approach, benefits could be shared in the case of legal and political activities that are "sponsored" by a financial partner. Imagine just for a moment a world in which global financial houses like Goldman Sachs devote their intellectual, financial, and political capital, not to the exploitation of dubious offset opportunities such as HFC-23 capture, but to the identification and promotion of critical sites of political intervention by disempowered voices for sustainability. Contrast such a world with the existing carbon markets' tendency "to encourage private corporations and technical experts to expend ingenuity on inventing novel, geographically far-flung market 'equivalents' for emissions reductions rather than finding ways to implement a structural shift away from fossil fuels."[59]

The carbon upset approach is targeted at challenging the political and economic inertia of the status quo. It seeks to introduce dynamism into our political economy by actively seeding disruption and potential transformative change. Conventional climate change policies such as carbon offsets and allowance giveaways have the perverse effect of further subsidizing already massively subsidized and politically dominant industries, thereby "entrenching institutions and procedures that are likely to stand in the way of constructive approaches to climate change."[60] Put bluntly, the main beneficiaries of the existing carbon offset system are the same industries and technologies that need to be radically transformed. Individuals and groups that are pursuing the transformation of such industries and technologies should be rewarded, and even encouraged into existence, through carbon upsets. Again, nothing in the counterfactual logic of carbon offsets prevents their use in this more dramatic and politically ambitious fashion. The limitation lies in our imagination.[61]

THE UNBEARABLE LIGHTNESS OF EMITTING

The offset system's tethering to neoliberal imagination becomes even more plain in the case of voluntary offsets. Unlike the compliance offset market, which exists to enable firms facing emissions caps to meet their regulatory obligations more cheaply, the voluntary offset market is targeted at individuals and organizations that wish to purchase credits to "offset" their emissions even in the absence of legal mandates. Greenhouse gas mitigation on this approach occurs through voluntary, decentralized choices generally made within the context of market transactions. Before completing an airline ticket purchase, for instance, passengers might be given the opportunity to offset their flight emissions for an additional fee. Before indulging in a porterhouse steak, a diner might be asked whether she wants to accept a surcharge so the restaurant could purchase emissions credits to offset her carbon-intensive meal choice.

On this approach the neoliberal order again is not threatened; rather, it is reinforced because the voluntary offset system, like the compliance offset system, obscures "the political nature of collective decisions."[62] Conveniently, agency and efficacy are seen to reside in individual economic choices in the market, rather than in mechanisms for political decision-making. The yawning chasm between individual actions and collective consequences—perhaps the defining feature of climate change as a global conundrum—is bridged through the conjuring magic of counterfactual carbon.

This bridge is illusory, for individuals are denied a way to imagine and realize an alternative world in which their choices do not contribute to climate change and therefore do not demand offsetting in the first place. *That* conversation is one many are aching to have, for our collective climate consciousness is finally awakening to the fact that we are, indeed, riding a bullet train to catastrophe. The salient templates for that conversation, though, remain largely neoliberal and inadequate. As a result, public discourse now features plenty of debate and hand-wringing regarding the individual ethics of having children, eating meat, traveling via airplane, owning a car, etc. But there is all too little conversation regarding the structure and regulation of food systems, electricity grids, public transportation networks, or any of the other technopolitical assemblages that determine the level of harm to be anticipated from individual decisions. To be sure, the emerging climate responsibility discourse does represent an advance over prior iterations, in which consumers have been offered convenient salves (e.g., change that light bulb, recycle that bottle) that are relatively disconnected from the major impacts of human choice and behavior. Still, the conversational template remains radically truncated.

Consider the decision whether to have children, which many now see as a profound moral question from the climate perspective. Under the current carbon market system, an individual concerned about the greenhouse gas emissions of her offspring would have to purchase offsets representing the expected additional contribution of the child and its generational descendants to the climate problem. According to one study, this would represent nearly 60 tons of CO_2 equivalent emissions per year.[63] Assuming a conservative carbon price of US$50 per ton, the conscientious parent would need to purchase US$3,000 per year of credits to offset the impact of deciding to have a child. Alternatively, in a different political economy, the voluntarily childless might be entitled to receive US$3,000 per year of credits for having chosen not to reproduce. The critical point is, whichever scheme is adopted, the environmental and regulatory consequences of the decision will be small in comparison to the impact that having or not having a child bears on one's life course.

Because we exist embedded in systems not of our choosing and well beyond our control, our greenhouse gas emissions have a certain unbearable lightness. They are light in the sense that our choices within systems give rise to emissions with little thought or means to avoid them; they are light in terms of the actual environmental impact they cause on their own, disaggregated from the emissions of billions of other individuals; and they are light in comparison to the economic costs or benefits that they might bring us within currently imaginable regulatory schemes. Yet, despite their lightness, they are unbearable when we countenance them as beings with agency and responsibility who wish to be ethical in the Anthropocene.

Carbon Superspreaders

Some greenhouse gas emissions appear to be even lighter than others. Carbon markets trade on the fact that, from the perspective of atmospheric chemistry, it does not matter where greenhouse gas emissions are reduced. Once dispersed within the troposphere, a ton of carbon is just a ton of carbon. From the perspectives of history and justice, however, the source and use of emissions matter a great deal. Henry Shue's distinction between "subsistence emissions" and "luxury emissions" nicely captures the idea that it matters both *who* is emitting and *for what purpose* they are emitting.[64] Disparities among nations in terms of historical, absolute, and per capita emissions levels mean that the climate change problem is rife with questions of responsibility and equity.[65]

Not only have nations contributed at vastly different levels to the climate change problem, but the problem itself inflicts harm in unequal ways. Poor countries, disproportionately concentrated around the equator, are more vulnerable to the impacts of increasing temperatures. Indeed, they *already* have suffered significantly: Researchers estimate that the gap in per capita income between the richest and the poorest countries in the world is 25% larger than it would have been in the absence of the human-induced climate change that occurred from 1961 to 2000.[66] Without the effects of anthropogenic climate change during that period, for instance, India would have been 30% wealthier and Nigeria 29% wealthier. Oil-rich Norway, on the other hand, has seen net *gains* from climate change as its high latitude climate has become more temperate.[67] A subsequent study coupled temperature-driven income losses with a warming attribution model to assign national shares of responsibility for the losses. Just five nations (the United States, China, Russia, Brazil, and India) were collectively responsible for US$6 trillion in income losses from warming since 1990, comparable to 11% of annual global gross domestic product.[68]

Climate justice typically is discussed in these country-to-country terms. Analysts have long recognized, however, that climate justice also raises questions of equity at the individual level.[69] Global disparities in individual contributions to the climate change problem are significant. By one estimate, the greenhouse gas emissions of someone in the wealthiest 1% of individuals in the world are 175 times as great as someone in the poorest 10%.[70] Between 1990 and 2015—a critical period in which global annual emissions rose by 60% and cumulative emissions since the beginning of the Industrial Revolution doubled—the richest 10% of the world's population were responsible for an estimated 52% of the cumulative emissions increase.[71] Similarly, the United Nations calculates that the top 10% of income earners in the world in 2015 were responsible for 48% of global CO_2 consumption emissions, whereas the bottom half of earners contributed only 7%. The top 1% of earners were responsible for 15% of global CO_2 consumption emissions, more than double the share of half the world's population.[72] Put another way, "just one-hundredth of the world population (77 million individuals) emits about 50% more than the entire bottom half of the population (3.8 billion individuals)."[73]

Analyses such as these prompted a leading academic research organization to conclude that the fundamental organizing principle of international climate negotiations—which remains largely focused on country-to-country comparisons as a basis for making equity and burden-allocation claims—no longer captures the empirical nuances of climate inequity:

> Carbon inequalities within countries now appear to be greater than carbon inequalities between countries. The consumption and investment patterns of a relatively small group of the population directly or indirectly contribute disproportionately to greenhouse gases. While cross-country emission inequalities remain sizeable,

overall inequality in global emissions is now mostly explained by within-country inequalities by some indicators.[74]

More precisely, whereas 62% of global carbon inequality in 1990 was explained by country-to-country differences, by 2019 within-country differences had risen to account for 64%.[75] Looking at the distribution of emissions by geographic region in figure 1.2 reveals that the conventional typology of developed and developing countries within international climate negotiations needs to be refined. Carbon superspreaders exist in all regions of the world, even the poorest.[76] Such trends appear poised to continue. Looking forward, the top 1% of earners are projected to have per capita consumption emissions levels in 2030 that are thirty times higher than the level that would be compatible with a 1.5°C pathway if emissions were equally distributed across the global population.[77] Strikingly, while household emission levels in the United States declined by an average 16% over the period from 1996 to 2019, emissions of the top 1% of households by income level increased by 23% and the top 0.1% by 50%.[78] That latter extremely rarified group each consumed an estimated 800 tons of CO_2 equivalent emissions per year (figure 1.3), an amount dramatically higher than the rest of

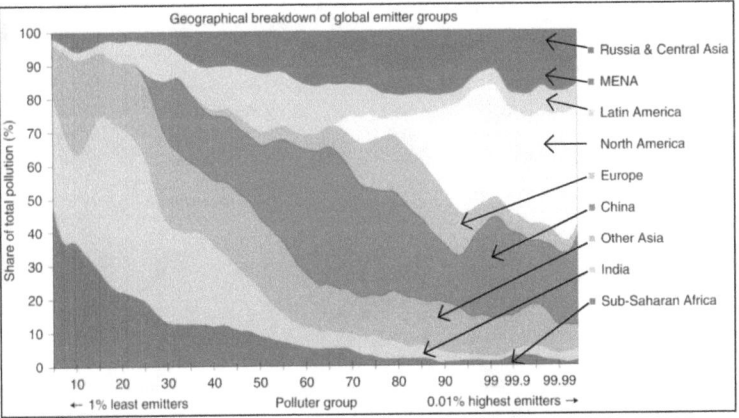

FIGURE 1.2. Geographical breakdown of global emitter groups. Source: Lucas Chancel, "Global Carbon Inequality over 1990–2019," *Nature Sustainability* 5 (2022): 931–938.

the 1% elite and radically out of line with the 2.3 tons per person level by 2030 estimated to be needed to keep the world on track to limit warming to 1.5°C above preindustrial levels.[79]

Even these jarring statistics can be topped if one looks more closely at the very top of the very top: An ingenious study of the carbon footprint of the world's ultrawealthy found that Russian oligarch Roman Abramovich's largest yacht, the *Eclipse*, by itself emitted more than 20,000 tons of CO_2 equivalent greenhouse gases in 2018.[80] That amount exceeded the annual emissions of the entire nation of Tuvalu, a low-lying Pacific island state that has begun archiving a virtual reality simulation of its territory in an effort to preserve the island's heritage and assert continued sovereign existence under international law even after it becomes uninhabitable due to sea level rise.[81] Scholars and advocates have increasingly targeted the luxury emissions of the world's wealthiest individuals and families as a focus of research and potential policy intervention.[82] A study in *Nature Climate Change*, for instance, reported results of a

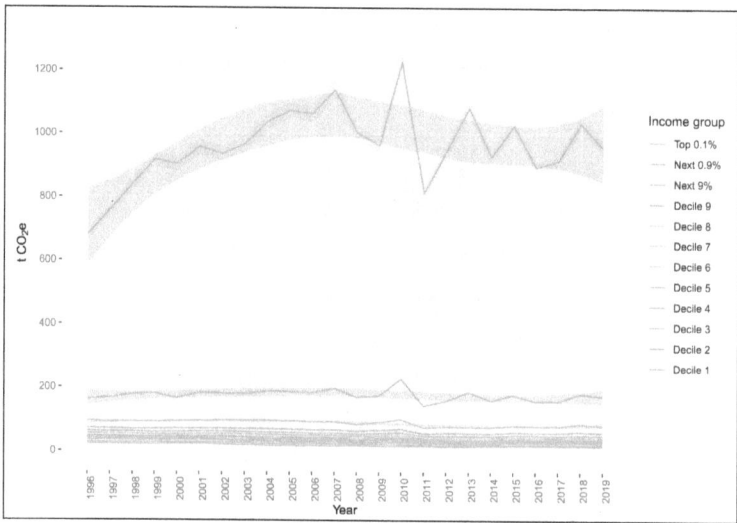

FIGURE 1.3. Mean US household carbon footprint per income group. Source: Starr et al., "Assessing U.S. Consumers' Carbon Footprints Reveals Outsized Impact of the Top 1%," *Ecological Economics* 205 (2023): 107698.

survey of high-net-worth individuals and their consumption habits with respect to private motor vehicles, air travel, household energy use, and spending on food and education. While not as massive as the emissions of a Russian oligarch, the study still concluded that "a typical superrich household of two people produces a carbon footprint of 129.3 tCO$_2$e per year," an amount around ten times the global per capita average.[83] A carbon footprint analysis of households within China found similar, though less pronounced disparities, with urban ultra-rich households comprising just 5% of the population being found responsible for 19% of greenhouse gas emissions from household consumption in all of China.[84]

Aviation is particularly instructive as an example of consumption inequality and the potential for effective climate policy intervention. The aviation sector is responsible for approximately 2.5% of emissions globally but drives a higher amount of warming due to various non-CO$_2$ effects.[85] It is also a sector whose services are utilized by a concentrated few in global terms. Researchers estimate that during 2018, only 11% of people around the world traveled by plane at all, with no more than 4% taking an international flight. Without even considering private air travel, a mere 1% of the global population was responsible for half of all commercial flight emissions that year.[86] Another study estimates that the top 10% of earners in the world consume around 75% of all energy for air transport, compared with only 5% used by the poorest 50%.[87] Even in the United States, researchers estimate that more than half of adults do not fly at all, while the 12% of adults who take six flights or more per year account for 68% of all flights taken in a given year.[88] In contrast to conventional carbon taxes, which pose significant regressivity challenges in practice, a tax on aviation travel could easily be designed in a graduated way to target well-off frequent flyers without imposing undue hardship on lower- or middle-income households.[89] Design of a yacht tax, one suspects, would be even easier.

The Efficiency of Equity

Related to the issue of luxury emissions is the question of whether inequality itself—as distinct from absolute levels of income and wealth—could be a driver of higher greenhouse gas emissions.

Research on the relationship between inequality and greenhouse gas emissions remains underdeveloped, although the issue has received more scholarly attention of late. Consistent with the notion that the marginal propensity to consume declines with income, an early study using data from forty-two countries between 1975 and 1992 found that income inequality is negatively associated with carbon emissions.[90] The authors interpreted their results to suggest that "a static trade off exists between reducing carbon emissions and promoting lower inequality both between and within countries."[91] Subsequent work has challenged this idea, finding a positive association between inequality and greenhouse gas emissions, especially but not only for higher-income countries and particularly in more recent decades as emissions levels have accelerated exponentially in most parts of the world.[92] An analysis of state-level data within the United States similarly found that emissions are positively associated with higher concentrations of income among the top 10% of a state's population.[93] In addition, a study focused on wealth rather than income inequality found a consistent positive relationship between wealth inequality and per capita greenhouse gas emissions in twenty-six high-income countries between 2000 and 2010.[94] Finally, a more recent analysis found that among thirty-five developed countries from 1985 to 2011, rising income inequality led to a tighter coupling between economic growth and CO_2 emissions, precisely the opposite of the decoupling that is desperately needed.[95]

A number of theoretical explanations have been proposed for a positive relationship between inequality and per capita greenhouse gas emissions, including: (1) inequality exacerbates the "power-weighted social decision rule," according to which the wealthy—who benefit more and suffer less from environmental degradation—have greater power to protect their interests in social decision-making processes by keeping capital free of regulatory burdens;[96] (2) inequality reduces social cohesion, cooperation, and trust, potentially inhibiting collective action to protect the environment;[97] (3) inequality impedes diffusion of green innovations to the mass market;[98] and (4) inequality leads to Veblen effects whereby consumption of energy-intensive goods and services increases as individuals aim to emulate and compete with the status-based consumption of the wealthy.[99] One recent comprehensive analysis of how "socioeconomic inequalities causally contribute to climate change" identified ten distinct drivers, including the preceding explanations as well as those focused on the role of inequality in impeding lower-income individuals and households from supporting climate policies.[100] The authors' summary of their analysis is worth quoting at length:

> Economic inequality creates power imbalances that enable capital interests to expand carbon-intensive production and obstruct climate policy, and it empowers the wealthy to live unsustainably carbon-intensive lifestyles that set standards of consumption to which those on lower incomes aspire. At the other end of the income distribution, poverty, un(der)employment, and financial insecurity leave people trepidatious of ambitious carbon-centric policies that threaten to erode their purchasing power or deprive them of decent work. These economic effects are often geographically clustered and layered onto existing spatial inequalities, fueling political backlash. Finally, it is plausible that social and economic inequalities undermine the social bonds of trust necessary for transformative collective climate action.[101]

Among these many drivers, certain consumption patterns of the ultrawealthy deserve special focus as evidence suggests they are disproportionately more destructive of the climate than those of poor people.[102] For instance, studies suggest that the elasticity of mobility-related consumption such as private vehicle use and air travel is greater than 1,[103] and that increasing housing wealth among the rich leads to more emissions than among the poor.[104] Given that private vehicle use, air travel, and housing selection are three of the most greenhouse gas–intensive consumption choices individuals can undertake, the same dollar in the hands of an ultrawealthy person may well cause more greenhouse gas emissions than a dollar in the hands of a poor person. Indeed, a comprehensive study of goods and services consumption among different income groups in eighty-six countries found income elasticity greater than 1 in several emissions-intensive categories such as vehicle fuel and packaged vacations.[105] In short, the poor may have a higher marginal propensity to consume in general, as one would expect, but the rich seem to have a higher propensity to consume in ways that make them carbon superspreaders.

These findings raise the possibility of viewing inequality through the lens of climate change mitigation.[106] Rather than seeing mitigation policy as posing separate dimensions of efficiency (e.g., how, and by how much, should emissions be reduced?) and equity (e.g., who should bear the costs of emissions reductions?), one can see the latter as offering a pathway to promote the former. This is because "lowering inequality will be good for the environment and decrease carbon dioxide levels in countries above a certain income threshold of GDP per capita."[107] Accordingly, limiting luxury emissions and moving financial resources from very rich people to poor people through measures such as steeply progressive income taxes, wealth taxes, or financial transaction taxes may be seen as a climate change *mitigation* measure and not merely as a policy concerning equitable redistribution.

As noted in the previous section, one also could target the carbon-intensive behaviors of the ultrawealthy directly through measures such as usage limits or consumption taxes on airline travel, vacation home ownership, meat consumption, and similarly impactful decisions.[108] To illustrate, a recent study modeled the effect of regulating energy use by high- and low-income households within twenty-seven European countries, finding that limiting usage from the top fifth at a relatively high level (specifically, leveling down all members of the top fifth to the eightieth percentile usage level) led to 9.7% overall reductions in greenhouse gas emissions from energy consumption, while increasing usage from the bottom fifth who live in poverty to a relatively low level (leveling up all members to the twentieth percentile level) increased emissions only by 1.4%.[109] Promoting greater equity in energy usage thus worked simultaneously to lower overall emissions.

The point here is not to analyze which is preferable as a mitigation instrument, i.e., targeting the wealthy directly or the carbon-intensive behaviors that tend to increase with wealth. The point is simply to note, as a conceptual matter, that limiting wealth and reducing inequality may offer underappreciated climate change *mitigation* benefits. Neoclassical economic theory typically assumes that promoting equity goals imposes a deadweight *loss* to society that is thought to result from distorting labor and investment incentives through the tax-and-transfer system. But the climate crisis shows there may be efficiency *gains* to society in the form of reduced luxury emissions when wealth is limited or is moved from the very rich to the poor or public funds. Just as law and economics scholars have begun to perceive the equity effects embedded within efforts to promote efficiency,[110] they should also attend to the efficiency effects of promoting equity. Such a perspective opens the possibility that "integrating certain carbon-centric policies into a wider program of social, economic, and democratic reforms would achieve decarbonization more effectively than carbon-centric policies alone."[111]

As another example of this reconceptualization, consider the fact that research indicates nations with greater female representation in parliament or other governing bodies have lower climate footprints controlling for other drivers.[112] One particularly sophisticated econometric study found that "female representation in national parliaments leads to more stringent climate change policies across countries, and by doing so, it results in lower carbon dioxide emissions."[113] Studies of female access to education and national measures of carbon intensity find a similar emissions-mitigating influence from greater female educational opportunity.[114] In a noteworthy longitudinal study using an index of women's political empowerment across numerous variables in seventy-two countries from 1971 to 2012, researchers found that CO_2 emissions decreased by about 11.5 percentage points in response to a one-unit increase in the empowerment index, leading the authors to conclude that "improving the status of women worldwide, especially in the developing countries, can reduce CO_2 emissions."[115]

In all these years of academic squabbling over carbon taxes and cap-and-trade, why have we not explored the possibility that dismantling political patriarchy might reduce emissions with essentially *no* legitimate social cost? Why are we instead subsidizing at up to $85 per ton untested technologies for sequestering carbon that merely perpetuate our dependency on a fossil fuel economy? Why not send $85 per ton to organizations that help women run for office? More to the point—why have we not run the serious economic analyses that would determine which is a more cost-effective way of reducing greenhouse gas emissions? Again, the point is not to assert these causal claims definitively, but rather to point out that low-cost and effective climate mitigation policy tools might be lying hidden in plain sight, if only our dominant ideological framework would allow us to recognize them.

On the other hand, to claim that these tools are lying hidden in plain sight may overstate the case. After all, the empirical possibilities considered in this section depend on the expertise of social scientists and other academic researchers, raising risks of technocratic inaccessibility similar to those that plague conventional climate economic concepts such as the social cost of carbon.[116] The fact that the policy tools discussed in this section address social goals that are independently understandable and desirable to many renders them perhaps more translatable into democratic discourse than the optimization models of conventional economists. Rather than trying to convert all social goals into a single welfare metric that is opaque to non-economists, the tactic simply identifies the reduction of greenhouse gas emissions as a co-benefit to goals that are already well articulated within familiar justice frameworks.

A separate concern might be that the experts relied upon in this section have simply "cherry-picked" progressive social goals for empirical study in order to connect them to climate change mitigation, rather than taking a more comprehensive survey of the many policies that might also have underappreciated climate co-benefits. Limiting refugee migration from developing to developed countries, for instance, might prevent emissions increases that would otherwise occur as migrants adapt to the higher per-capita emissions lifestyles that prevail within developed countries. The point here is not to catalogue the variety of alternative policy approaches to mitigation that might be pursued, progressive or nonprogressive. The point instead is to emphasize the need in general to address decarbonization with a fuller moral and political imagination than we are permitted to do through the constraining lens of climate economics. What is demanded is fulsome democratic engagement with the variety of ways in which societal transformation can be achieved.

Out of Scale

Viewing the reduction of wealth and gender inequality as being, in part, climate mitigation measures might help more broadly to destabilize the neoliberal frame of policymaking. Neoclassical economics' conceptual separation of allocative efficiency and just distribution indirectly preferences the current distribution of endowments and power in a society because any change from that distribution appears "inefficient" on the allocation criterion. The government may only have a "leaky bucket" with which to *re*-distribute,[117] but the government also sets policies that determine *initial* distributions and the resulting amount of so-called *re*-distribution that is required by principles of justice.[118] Initial distributive judgments should be thought of as market-determining, not market-determined. Thus, social decisions regarding certain aspects of distribution—such as determining who may benefit from the atmosphere's finite ability to absorb greenhouse gases—should occur separate from and prior to the reallocation of resources through the market or through use of the tax-and-transfer to address market injustices. Proponents of neoclassical economics could *then* practice their bread and butter of maximizing the market's allocative efficiency, but only after the optimal scale and distributive foundation of that market have first been addressed.

The neoliberal paradigm tends not to recognize environmental sustainability as an independent policy goal. Instead, sustainability is subsumed within efficiency with the claim that, once government addresses negative environmental externalities and "gets the prices right," markets will efficiently allocate all resources including environmental ones across both space and time. Alyssa Battistoni's contribution to this volume powerfully explores the rise of this simplistic market-oriented viewpoint, in which even "a phenomenon described as an 'existential threat' to humanity has been reduced to a problem of missing prices."[119] As Battistoni reveals, the viewpoint obscures deep questions of power, agency, and responsibility that drive the realization, scope, and magnitude of "externalities." Neither minor nor unintentional, externalities instead should be seen as part of the fundamental logic of market economies.

For empirical support of this notion, consider a recent analysis by leading economists regarding the scale of adverse climate impacts from nearly 15,000 publicly traded global corporations.[120] Using the EPA's then-prevailing social cost of carbon of US$190 per ton and focusing conservatively just on corporations' direct emissions from owned or controlled sources (i.e., Scope 1 emissions), the researchers found that firms presently are externalizing climate harms equal to around 44% of their operating profits. This does not mean that "internalizing" those carbon damages through a tax of US$190 per ton would immediately reduce those firms' profits by almost half—calculating the incidence of such a policy is a complicated task. But it does imply that the negative externalities of the modern economy are of such a magnitude that they no longer should be considered "external." Indeed, for some countries and sectors, the ratio of externalized climate costs to operating profits appears to be the very heart of the business model. For instance, Russian companies analyzed by the researchers impose climate harms at nearly 130% the level of their operating profits. Within an individual, such agents would be called a cancer.

It is time to return to fundamentals: Is the human economy a subsystem of the environment, or is the environment a subsystem of the economy? The former vision emphasizes natural constraints on the expansion of human production, including both the scarcity of resource inputs to the economic process and the scarcity of pollution sinks to absorb waste outputs of the process. The latter vision admits of no such limits to human economic growth, given that no conceptual superstructure such as the environment exists "around" the economy to constrain it. Economic growth is limited only by the availability of human-made capital and labor, not by natural resources. This dematerialized economy is made possible by production functions in which natural and human capital are treated as infinitely substitutable. Thus, although specific types of natural inputs can in fact become scarce or depleted, no general scarcity of natural resources can ever constrain economic growth.[121] In an inspired and revealing contrast to this orthodoxy, two researchers recently developed a novel climate integrated assessment model that includes "natural capital" as a distinct form of wealth that can become undermined by the impacts of climate change. In stark contrast to conventional models such as DICE, these researchers found that the economically "optimal" climate pathway limits the global average surface temperature increase to 1.5°C by 2100.[122] By incorporating in an analytically rigorous way humanity's dependence on the environment, the researchers offered a dramatic advance over DICE and other conventional models, which assume that as much as 90% of economic activity is immune from negative climate impacts simply because it takes place indoors.[123]

This elementary distinction in pre-analytic vision leads to dramatically different orientations to policymaking. As pioneering ecological economist Herman Daly put it: "When we draw a containing boundary of the environment around the economy, we move from 'empty-world' economics to 'full-world' economics. Economic logic stays the same, but the perceived pattern of scarcity changes radically and policies must change radically."[124] Particularly, the goal of market regulation becomes far more complicated than merely seeking to maximize allocative efficiency while taking most all other social and natural conditions as given. More important than managing market conditions to allow resources to be devoted to their most-valued use, governments must manage the absolute scale of the human macroeconomy to ensure that its impact does not exceed the carrying capacity of the environment. "Scale" refers to the material impact of the economy in relation to the natural environment. More precisely, it is a measure of "the physical volume of the throughput, the flow of matter-energy from the environment as low-entropy raw materials and back to the environment as high-entropy wastes."[125]

In addition to managing scale, governments also must attend more vigorously to the distribution of wealth. Outside of the neoliberal imagination, environmental crises and social inequities can no longer be ignored on the grounds that nature and opportunity are both infinite in their bounty. Such limitless visions have always lacked a defensible scientific foundation; in the climate century, unmistakably so.[126] Thus, once one accepts that the level of material throughput in the economy is subject to biophysical constraints such as the atmosphere's ability to receive greenhouse gas emissions without fundamental disruption, then the concerns of sustainability and distribution can no longer be pushed to the side in ever-intensifying pursuit of allocative efficiency.[127] Rising tides may lift boats, but they also flood coastlines, destroy wetlands, and salinize water supplies, and they do none of these things fairly.

As with the social cost of carbon, essential work is done in welfare economics at the level of assumption. Neoliberal approaches to environmental policymaking take as given nonmaterial factors such as consumer preferences, income distribution, resource rights, and societal desire for equity. The economist's task becomes simply to ensure that material factors are most efficiently deployed to suit the given nonmaterial parameters. In practice, the adjustment of material factors almost always involves economic growth created by ever-intensifying exploitation of natural and human resources. That same expansion in economic output provides the basis for political assurances that economic inequality will somehow be alleviated through the undirected rising tide. In the climate century, by contrast, one must take the physical environment as posing fixed constraints and instead contemplate mechanisms for adjusting nonmaterial factors to best suit the physical world within which we inescapably reside.[128] That is, one must study "how the nonphysical variables of technology, preferences, distribution, and lifestyles can be brought into feasible and just equilibrium with the complex biophysical system of which we are a part."[129]

Today, Daly's call resonates with the Green New Deal and degrowth movements,[130] both of which can be understood as being concerned not merely with satisfying existing individual wants, but also with encouraging new cultural modes of relating to the environment and of defining a life well-lived and a just society to live it in. The urgency of such alternative economic ideologies is underscored by recent research on the woeful inadequacy of current rates of decoupling between economic growth and consumption-based greenhouse gas emissions.[131] Examining eleven out of thirty-six high-income nations that managed to achieve some amount of absolute decoupling between 2013 and 2019, the researchers found that the eleven countries would require more than 220 years to reduce their emissions by 95%, emitting in the process twenty-seven times their allocated equitable share of the remaining 1.5°C-compliant carbon budget (with equitable shares being estimated conservatively based on population size alone without regard to historical responsibility, capacity, or other equity-relevant factors). To meet the countries' 1.5°C-compliant reduction burden, decoupling rates would need to increase by a factor of ten before 2025. In words that echo Daly, the researchers concluded by arguing that, going forward, "[a] crucial step is to stop the pursuit of aggregate economic growth and instead pursue post-growth approaches oriented towards sufficiency, equity, and wellbeing."[132]

Daly's words also resonate with Tribe's warning in "Ways Not To Think About Plastic Trees" that an unduly narrow welfare economic lens for environmental policymaking might unintentionally alter affected individuals' "preferences, attitudes, and cost-benefit conceptions over time, and hence their character as a society of persons interacting with one another and with the natural order."[133] Rather than recognize that the highest human rationality is to reason about the things we want to want—not merely reason about how to get what we (assume we) want[134]—the neoliberal policy approach aims simply to maximize existing preferences, however crudely they may be apprehended and irrespective of whether their content implies a pathway to peril. Failure to heed the warnings of scholars like Daly and Tribe has led not only to decades of continued emissions growth, but also to the naturalization of a worldview in which such growth is absurdly deemed welfare-maximizing. Indeed, in his Nobel address, Nordhaus continued to report such dangerously flawed DICE findings as "damages are estimated to be 2 percent of [global economic] output at a 3°C global warming and 8 percent of output with 6°C warming," and that "the cost-benefit optimum rises to over 3°C in 2100."[135] The world in which climate damages are somehow limited to 8% of output at 6°C warming, or in which social welfare is maximized at over 3°C warming, is a model world that existed first in Nordhaus's imagination but then spread to become a convenient fantasy for us all. All models are wrong, but some are dangerous.

Conclusion

Tribe's seminal contribution to environmental law scholarship carries an awkward postscript. In recent years, Tribe has become better known for his representation of the coal industry and other fossil fuel interests in their ardent opposition to climate change regulation and common law accountability. How do we make sense of the fact that the author of "Ways Not To Think About Plastic Trees" devoted his considerable talent and reputation to representing Peabody Energy and the American Petroleum Institute in their efforts to block, respectively, President Obama's Clean Power Plan and the ability of state courts to entertain claims that fossil fuel producers have substantially contributed to the public nuisance of climate change? Popular reaction to Tribe's fossil fuel advocacy was harsh, with headlines asking questions such as, "Did Laurence Tribe Sell Out?," "Et tu Tribe?," and "Did Laurence Tribe Sell His Soul to

Big Coal?" In response, Tribe argued that, although he has long represented an array of clients for compensation, he has only done so when he is able to advance a position that he personally believes in. Indeed, in the case of the Clean Power Plan, Tribe deliberately injected his personal professional identity into the matter by submitting comments to the EPA not just on behalf of his client, Peabody, but also on behalf of "Laurence H. Tribe."

This personal vouching led some to speculate that Tribe might be suffering a sort of ideological capture: Having represented corporate interests with increasing frequency over the years, the famed scholar may have come to see arguments made on his clients' behalf as more compelling and reasonable through a sort of affiliation bias.[136] In other words, Tribe's views may have been altered by the very kind of endogenous cultural influence that the scholar worried over in "Ways Not To Think About Plastic Trees" with respect to environmentalists who operate within the welfare economic policy framework.

On his retirement, Tribe gave an interview to the Harvard alumni magazine in which he expressed regret over his work for Peabody:

> When I agreed to argue against Obama's clean-power plan I thought that it was illegal. I still think it probably was. I think it was a stretch of the Clean Air Act. But in hindsight, because of the existential importance of dealing with climate change, I think that's a case where I should've basically looked the other way, not let my legal theories drive me to the conclusion they did.[137]

Tribe's recognition of "the existential importance of dealing with climate change" in the interview stands in tension with a line that he regularly offered in opposition to the Clean Power Plan during his Peabody engagement: "I don't think burning the Constitution instead of coal will really be a way of saving the environment."[138] The US Constitution was written at a time when the world contained less than one billion people and the Industrial Revolution had hardly begun. It is not surprising that the document failed to contemplate the governance challenges posed by climate change. Whether the Constitution constitutes a suicide pact with respect to climate change, however, is up to the scholars, parties, advocates, and judges whose actions today drive constitutional interpretation and lawmaking.

Even more germane to this chapter is Tribe's hindsight wish not to have let his "legal theories drive [him] to the conclusion they did." This chapter has argued that economics also can be a theory that drives regrettable conclusions when it comes to climate change. Rooted in neoliberal imagination and claiming to offer the only practical policy response to climate change, promoters of the welfare economic approach promise to maximize human wellbeing by identifying an emissions pathway that balances the negative impacts of greenhouse gas emissions against the cost of abating them, and by offering market-based policy instruments such as tradeable carbon emissions permits to promote that pathway at least cost. As this chapter has argued, however, the welfare economic approach is riddled with limitations and confusions that threaten to misstate in catastrophic ways what is considered desirable and possible in the realm of climate action.

Still, recognizing the framework's continuing purchase, the chapter has described two thought projects that are locatable within the economic framework and that hold potential to open space for new moral and political imaginings to take root. First, the carbon upset approach was offered as an exercise in expansive thinking about how to incentivize greenhouse gas emissions reductions. The upset approach uses the same counterfactual reasoning as carbon offsets, but admits into consideration the role of activism, law, and politics in reducing emissions, not just finance and development. By enriching and empowering agents that stand diametrically opposed to the entrenchment of the fossil fuel economy, the carbon upset approach offers a glimpse into a worldview that does not embrace calamity as an inevitable side effect of the current distribution of political and economic power. Instead, the carbon economy would be turned against itself so that a new political economy might eventually emerge.

Second, limiting wealth and reducing inequality were explored as greenhouse gas mitigation options. Over the past three decades of concerted international climate policy discussion, economists have consistently insisted that instruments such as the social cost of carbon and carbon offsets should be utilized to achieve least-cost emissions reductions. It is curious that limiting wealth and reducing inequality have not been considered more seriously as means of limiting emissions. Within mainstream economic theory, such measures are generally considered to impose a social welfare *loss* in the form of distorted incentives that must be weighed against whatever equity goals are achieved. But the enormous climate externalities of affluence suggest there may also be social welfare *gains* to be had from making the world a more equal place. Most striking about this example conceptually is the fact that inequality reduction is almost never discussed as a mitigation or "efficiency" measure in climate policy, notwithstanding its potential to reduce emissions at a cost that may be lower than heavily studied and subsidized alternatives such as carbon capture and storage. Our current "ways of seeing" in environmental law may be simply too narrow to perceive such possibilities.[139]

Neither of these thought projects would be equal to the challenge of ensuring a just and sustainable economy. But we must start from the mountain we are on, even as we seek new ones to climb.

Acknowledgments

For invaluable research assistance, the author thanks Sarah Baldinger, Kyle Bigley, and Yash Chauhan. For helpful comments and feedback, the author thanks William Boyd, Chiara Cordelli, Dale Jamieson, Fergus Green, Sheila Jasanoff, Alvin Klevorick, Matthew Kotchen, Melissa Lane, Kian Mintz-Woo, Nick Nugent, Zeynep Pamuk, Stefan Schaeffer, Kristen Stilt, Katrina Wyman, and participants of the American Society for Political and Legal Philosophy 2023 conference, a faculty workshop at Yale Law School, the Brooks Institute Annual Academic Workshop for Scholars, and a conference held on April 5–6, 2019 by the Program on Science, Technology and Society at the Harvard Kennedy School, "Multiple Carbons: Historical and Contemporary Approaches to Governance."

Notes

1 Laurence H. Tribe, "Ways Not To Think About Plastic Trees: New Foundations for Environmental Law," *Yale Law Journal* 83, no. 7 (1974): 1315–1348. Tribe's famous law review article emerged from a multiyear National Science Foundation project titled "Research Needs Concerning

the Incorporation of Human Values into Environmental Decision Making." The project was jointly sponsored by the American Academy of Arts & Sciences and the Center for Environmental Study at Princeton University, where it was led by Murray Gell-Mann and Robert Socolow, among others.

2 Tribe, "Ways Not To Think About Plastic Trees," 1317.

3 Tribe, "Ways Not To Think About Plastic Trees," 1330–1331.

4 For readers who are aware of Tribe's late-career representation of fossil fuel companies in opposition to climate action and who wish to see that apparent dissonance addressed, see for example text accompanying notes 136–138.

5 Cass R. Sunstein, *The Cost-Benefit State: The Future of Regulatory Protection* (Chicago: American Bar Association, 2003).

6 James Gustave Speth, *They Knew: The US Government's Fifty-Year Role in Causing the Climate Crisis* (Cambridge, MA: MIT Press, 2022).

7 Stephen J. DeCanio, *Economic Models of Climate Change: A Critique* (New York: Palgrave Macmillan, 2003), 2–4.

8 Anil Agarwal and Sunita Narain, *Global Warming in an Unequal World: A Case of Environmental Colonialism* (New Delhi: Center for Science and Environment, 1991).

9 Steffen Böhm, Maria Ceci Misoczky, and Sandra Moog, "Greening Capitalism? A Marxist Critique of Carbon Markets," *Organization Studies* 33, no. 11 (2012): 1617–1638.

10 William Boyd, "The Poverty of Theory: Public Problems, Instrument Choice, and the Climate Emergency," *Columbia Journal of Environmental Law* 46, no. 2 (2021): 399–487. It is humbling to read Dale Jamieson's prescient and powerful critique from his 1989 address to the American Association for the Advancement of Science, which covers similar conceptual terrain as this article: Dale Jamieson, "Ethics, Public Policy, and Global Warming," *Science, Technology, and Human Values* 17, no. 2 (1992): 139–153. The long-standing but nondecisive existence of such critiques raises a question whether climate economics really has been persuasive and influential *as an idea*, or whether instead its influence has been driven by the fact it often produces policy guidance that accords with powerful interests supportive of a fossil fuel economy. The latter possibility would help explain why the more neoliberal and conservative strands of climate economic thought seem to have held disproportionate influence over climate policy in the United States, despite the existence of numerous alternative approaches that would support more aggressive climate action.

11 Tribe, "Ways Not To Think About Plastic Trees," 1324.

12 Dale Jamieson, *Reason in a Dark Time* (New York: Oxford University Press, 2014).

13 Tribe, "Ways Not To Think About Plastic Trees," 1335.
14 Douglas A. Kysar, *Regulating from Nowhere: Environmental Law and the Search for Objectivity* (New Haven, CT: Yale University Press, 2010).
15 Douglas A. Kysar, "Politics by Other Meanings: A Comment on 'Retaking Rationality Two Years Later,'" *Houston Law Review* 48, no. 1 (2010): 43–77.
16 Cf. Richard L. Revesz and Michael A. Livermore, *Retaking Rationality: How Cost Benefit Analysis Can Better Protect the Environment and Our Health* (Oxford: Oxford University Press, 2008), 10, suggesting a pro-environment strategy of "mending, not ending cost-benefit analysis."
17 "Memorandum of January 30, 2009: Regulatory Review: Memorandum for the Heads of Executive Departments and Agencies," *Code of Federal Regulations*, title 3 (2010): 343–344. www.govinfo.gov.
18 Marlowe Hood, "Climate Economics Nobel May Do More Harm than Good" (July 6, 2020). https://phys.org.
19 William D. Nordhaus and Joseph Boyer, *Warming the World: Economics Models of Global Warming* (Cambridge, MA: MIT Press, 2003).
20 See, e.g., Interagency Working Group on Social Cost of Carbon, US Government, "Social Cost of Carbon for Regulatory Impact Analysis Under Executive Order 12866," Technical Support Document, Washington DC: 4: "In this context, statements recognizing the limitations of the analysis and calling for further research take on exceptional significance. The interagency group offers the new [social cost of carbon] values with all due humility about the uncertainties embedded therein and with a sincere promise to continue to work to improve them." For similarly candid and admirable explorations of the moral and political assumptions used in devising the social cost of carbon, see Kian Mintz-Woo, "A Philosopher's Guide to Discounting," in *Philosophy and Climate Change*, ed. Mark Bryant Budolfson, Tristram McPherson, and David Plunkett (Oxford: Oxford University Press, 2021), 90–110; and Marc Fleurbaey, Maddalena Ferranna, Mark Budolfson, Francis Dennig, Kian Mintz-Woo, Robert Socolow, Dean Spears, and Stéphane Zuber, "The Social Cost of Carbon: Valuing Inequality, Risk, and Population for Climate Policy," *The Monist* 102, no. 1 (2019): 84–109.
21 Steve Keen, "The Appallingly Bad Neoclassical Economics of Climate Change," *Globalizations* 18, no. 7 (2021): 1149–1177; Robert S. Pindyck, "Climate Change Policy: What Do the Models Tell Us?," *Journal of Economic Literature* 51, no. 3 (2013): 860–872.
22 Center for Biological Diversity v. National Highway Traffic Safety Administration, 538 F.3d 1172 (9th Cir. 2008).
23 Jamieson, "Ethics, Public Policy, and Global Warming."
24 David I. Armstrong McKay, Arie Staal, Jesse F. Abrams, Ricarda Winkelmann, Boris Sakschewski, Sina Loriani, Ingo Fetzer, Sarah E. Cor-

nell, Johan Rockström, and Timothy M. Lenton, "Exceeding 1.5°C Global Warming Could Trigger Multiple Climate Tipping Points," *Science* 377, no. 6611 (2022): eabn7950; Will Steffen, Johan Rockström, Katherine Richardson, Timothy M. Lenton, Carl Folke, Diana Liverman, Colin P. Summerhayes et al., "Trajectories of the Earth System in the Anthropocene," *Proceedings of the National Academy of Sciences* 115, no. 33 (2018): 8252–8259; Timothy M. Lenton, Hermann Held, Elmar Kriegler, Jim W. Hall, Wolfgang Lucht, Stefan Rahmstorf, and Hans Joachim Schellnhuber, "Tipping Elements in the Earth's Climate System," *Proceedings of the National Academy of Sciences* 105, no. 6 (2008): 1786–1793.

25 Interagency Working Group, "Social Cost of Carbon for Regulatory Impact Analysis Under Executive Order 12866," 2.

26 Zachary Liscow and Quentin Karpilow, "Innovation Snowballing and Climate Law," *Washington University Law Review* 95 no. 2 (2017): 387–464.

27 The possibility of reciprocal climate action by other countries has been used as an argument by academics and policymakers in favor of counting global, as opposed to merely domestic, impacts when calculating the social cost of carbon. See William Pizer, Matthew Adler, Joseph Aldy, David Anthoff, Maureen Cropper, Kenneth Gillingham, Michael Greenstone et al., "Environmental Economics. Using and Improving the Social Cost of Carbon," *Science* 346, no. 6214 (2014):1189–1190, and Peter Howard and Jason Schwartz, "Think Global: International Reciprocity as Justification for a Global Social Cost of Carbon," *Columbia Journal of Environmental Law* 42(S) (2017): 203–294. One analysis even goes so far as to argue that the ratio of the US emissions reduction commitment to the aggregate commitments by other countries under the Paris Agreement is roughly equal to the ratio implied by incorporating extraterritorial damages into the social cost of carbon, thereby justifying use of the global damages estimate. See Trevor Houser and Kate Larsen, "Calculating the Climate Reciprocity Ratio for the US," *Rhodium Group* (2021), https://rhg.com. Instrumentally valuable though they may be, these arguments remain strategically focused on national self-interest rather than on the underlying moral question of whether US policymakers and the constituents they represent should care about the impact of their choices and actions on residents of other countries.

28 Interagency Working Group, "Social Cost of Carbon for Regulatory Impact Analysis Under Executive Order 12866," 17.

29 Interagency Working Group, "Social Cost of Carbon for Regulatory Impact Analysis Under Executive Order 12866," 19.

30 For surveys that reach contrasting conclusions, see Douglas A. Kysar, "Discounting . . . On Stilts," *University of Chicago Law Review* 74, no. 1 (2007): 119–138, and Kian Mintz-Woo, "A Philosopher's Guide to Discounting."

31 Joshua Farley, "The Role of Prices in Conserving Critical Natural Capital," *Conservation Biology* 22, no. 6 (2008): 1399–1408.

32 Frank Ackerman, Elizabeth A. Stanton, and Ramón Bueno, "Fat Tails, Exponents, Extreme Uncertainty: Simulating Catastrophe in DICE," *Ecological Economics* 69, no. 8 (2010): 1657–1665.

33 Steven C. Sherwood and Matthew Huber, "An Adaptability Limit to Climate Change Due to Heat Stress," *Proceedings of the National Academy of Sciences* 107, no. 21 (2010): 9552–9555.

34 Nicholas Stern, "The Structure of Economic Modeling of the Potential Impacts of Climate Change: Grafting Gross Underestimation of Risk onto Already Narrow Science Models," *Journal of Economic Literature* 51, no. 3 (2013): 838–859.

35 Martin L. Weitzman, "Fat Tails and the Social Cost of Carbon," *American Economic Review* 104, no. 5 (2014): 544–546.

36 US Government Accountability Office, "Social Cost of Carbon: Identifying a Federal Entity to Address the National Academies' Recommendations Could Strengthen Regulatory Analysis" (2020). Accessed October 25, 2020. www.gao.gov.

37 In some cases, federal judges called the administration's bluff and ruled the attempted rollbacks unlawful based on the inadequacy of the administrative record regarding impacts. See, e.g., California v. Bernhardt, 472 F.Supp.3d 573 (N.D. Cal 2020). In such a case, the requirement that agencies produce a substantial cost-benefit analysis in support of their proposed actions can work in a salutary way to block attempted *de*-regulatory initiatives. Whether the requirement works in a salutary fashion for *pro*-regulatory aims is less clear.

38 National Academies of Sciences, Engineering, and Medicine, *Valuing Climate Damages: Updating Estimation of the Social Cost of Carbon Dioxide* (Washington, DC: National Academies Press, 2017).

39 See, e.g., Kevin Rennert, Frank Errickson, Brian C. Prest, Lisa Rennels, Richard G. Newell, William Pizer, Cora Kingdon et al., "Comprehensive Evidence Implies a Higher Social Cost of CO2," *Nature* 610 (2022): 687–692; Nicholas Stern, "A Time for Action on Climate Change and a Time for Change in Economics," *Economic Journal* 132, no. 644 (2022): 1259–1289; Gernot Wagner, David Anthoff, Maureen Cropper, Simon Dietz, Kenneth T. Gillingham, Ben Groom, J. Paul Kelleher, Frances C. Moore, and James H. Stock, "Eight Priorities for Calculating the Social Cost of Carbon," *Nature* 590 (2021): 548–550; and Tamma Carleton and

Michael Greenstone, "Updating the United States Government's Social Cost of Carbon," BFI Working Paper No. 2021-04, Becker Friedman Institute (2021). https://bfi.uchicago.edu.

40 Carleton and Greenstone, "Updating the United States Government's Social Cost of Carbon."

41 United States Environmental Protection Agency, "EPA Supplementary Material for the RIA for the Supplemental Proposed Rulemaking, NSPS and EG for Existing Sources: Oil and Natural Gas Sector Climate Review—EPA External Review Draft of Report on the Social Cost of Greenhouse Gases: Estimates Incorporating Recent Scientific Advances" (September 2022). Docket Number: EPA-HQ-OAR-2021-0317-1549. www.regulations.gov.

42 Interagency Working Group, "Social Cost of Carbon for Regulatory Impact Analysis Under Executive Order 12866," n. 6.

43 Rebecca Hersher, "Why the EPA Puts a Higher Value on Rich Lives Lost to Climate Change," *National Public Radio* (February 8, 2023). www.npr.org.

44 Kysar, "Politics by Other Meanings."

45 Larry Lohmann, "Marketing and Making Carbon Dumps: Commodification, Calculation and Counterfactuals in Climate Change Mitigation," *Science as Culture* 14, no. 3 (2005): 217.

46 Rebecca Lave, Martin Doyle, and Morgan Robertson, "Privatizing Stream Restoration in the US," *Social Studies of Science* 40, no. 5 (2010): 677–703.

47 Véra Ehrenstein and Fabian Muniesa, "The Conditional Sink: Counterfactual Display in the Valuation of a Carbon Offsetting Reforestation Project," *Valuation Studies* 1, no. 2 (2013): 162.

48 Bruno Latour, *Science in Action: How to Follow Scientists and Engineers through Society* (Cambridge, MA: Harvard University Press, 1987).

49 Lambert Schneider, "Assessing the Additionality of CDM Projects: Practical Experiences and Lessons Learned," *Climate Policy* 9, no. 3 (2009): 242-254.

50 Michael Wara, "Is the Global Carbon Market Working?," *Nature* 445 (2007): 595–596.

51 Barbara K. Haya, Samuel Evans, Letty Brown, Jacob Bukoski, Van Butsic, Bodie Cabiyo, Rory Jacobson, Amber Kerr, Matthew Potts, and Daniel L. Sanchez, "Comprehensive Review of Carbon Quantification by Improved Forest Management Offset Protocols," *Frontiers in Forests and Global Change* 6 (2023): 958879; Shane R. Coffield, Cassandra D. Vo, Jonathan A. Wang, Grayson Badgley, Michael L. Goulden, Danny Cullenward, William R. L. Anderegg, and James T. Randerson, "Using Remote Sensing to Quantify the Additional Climate Benefits of

California Forest Carbon Offset Projects," *Global Change Biology* 28, no. 22 (2022): 6789–6806.

52 Patrick Greenfield, "Revealed: More Than 90% of Rainforest Carbon Offsets by Biggest Certifier Are Worthless, Analysis Shows," *The Guardian* (January 18, 2023). www.theguardian.com.

53 Böhm, Misoczky, and Moog, "Greening Capitalism?," 1.

54 Mark Schapiro, "Conning the Climate: Inside the Carbon-Trading Shell Game," *Harper's Magazine* (February 2010): 38. https://harpers.org.

55 Lohmann, "Marketing and Making Carbon Dumps," 218.

56 Ehrenstein and Muniesa, "The Conditional Sink," 181.

57 Douglas A. Kysar, "Not Carbon Offsets, But Carbon Upsets," *The Guardian* (August 29, 2010). Accessed October 25, 2020. www.theguardian.com.

58 Lohmann, "Marketing and Making Carbon Dumps," 214.

59 Larry Lohmann, "Neoliberalism and the Calculable World: The Rise of Carbon Trading," in *Upsetting the Offset: The Political Economy of Carbon Markets*, ed. Steffen Böhm and Siddhartha Dabhi (London: MayFlyBooks), 25–41.

60 Lohmann, "Marketing and Making Carbon Dumps," 204.

61 It is fitting, then, that the only real-world instantiation of the carbon upset concept the author is aware of takes the form of a digital art exhibition launched in 2023. As described by the artists, their "platform includes a registry of alternative offsets that focus on social exchanges and political actions in order to contribute to a program of highly financialized radical change." See "About," *Offset*. https://offset.labr.io/about/. Accessed May 28, 2025.

62 Ehrenstein and Muniesa, "The Conditional Sink," 174.

63 Seth Wynes and Kimberly A. Nicholas, "The Climate Mitigation Gap: Education and Government Recommendations Miss the Most Effective Individual Actions," *Environmental Research Letters* 12, no. 7 (2017): 074024.

64 Henry Shue, "Subsistence Emissions and Luxury Emissions," *Law & Policy* 15, no. 1 (1993): 39–60.

65 Maxine Burkett, "Behind the Veil: Climate Migration, Regime Shift, and a New Theory of Justice," *Harvard Civil Rights-Civil Liberties Law Review* 53 (2018): 445–493.

66 Noah S. Diffenbaugh and Marshall Burke, "Global Warming Has Increased Global Economic Inequality," *Proceedings of the National Academy of Sciences* 116, no. 20 (2019): 9808–9813.

67 Somini Sengupta, "Global Wealth Gap Would Be Smaller Today Without Climate Change, Study Finds," *New York Times* (April 22, 2019): www.nytimes.com.

68 Christopher W. Callahan and Justin S. Mankin, "National Attribution of Historical Climate Damages," *Climatic Change* 172, no. 40 (2022): 1–19.

69 See for example Shoibal Chakravarty, Ananth Chikkatur, Heleen de Coninck, Stephen Pacala, Robert Socolow, and Massimo Tavoni, "Sharing Global CO2 Emission Reductions Among One Billion High Emitters," *Proceedings of the National Academy of Sciences* 106, no. 29 (2009): 11884–11888; Paul Baer, Sivan Kartha, Tom Athanasiou, and Eric Kemp-Benedict, "The Greenhouse Development Rights Framework: Drawing Attention to Inequality within Nations in the Global Climate Policy Debate," *Development and Change* 40, no. 6 (2009): 1121–1138.

70 "Extreme Carbon Inequality," *Oxfam*, 2015, www-cdn.oxfam.org. Accessed October 25, 2020.

71 "The Carbon Inequality Era," *Oxfam*, 2020, https://oxfamilibrary.openrepository.com. Accessed October 25, 2020.

72 United Nations Environment Programme, *Emissions Gap Report 2020* (Nairobi, 2020). The carbon footprint of the top 10% and 1% of households would be even higher if calculations included emissions attributed to investment-income generating activities, in addition to personal consumption. See Jared Starr, Craig Nicolson, Michael Ash, Ezra M. Markowitz, and Daniel Moran, "Income-Based U.S. Household Carbon Footprints (1990–2019) Offer New Insights on Emissions Inequality and Climate Finance," *PLOS Climate* 2, no. 8 (2023): e0000190.

73 Lucas Chancel, *Unsustainable Inequalities: Social Justice and the Environment*, trans. Malcolm DeBevoise (Cambridge, MA: Harvard University Press, 2020).

74 Lucas Chancel, Philipp Bothe, and Tancrède Voituriez, *Climate Inequality Report 2023* (World Inequality Lab Study, 2023), 5.

75 Lucas Chancel, "Global Carbon Inequality over 1990–2019," *Nature Sustainability* 5 (2022): 931–938.

76 Additional aspects of this figure bear noting, such as the substantial concentration of low-emitting individuals in India and sub-Saharan Africa, the surprising amount of high- and ultra-high-emitting individuals in China, the concentration of European individuals within relatively tight bands of medium to above-average emissions, and the extraordinary right-ward concentration of low-, medium-, and high-income individuals in the United States. As the author notes, "[i]t is striking that the poorest half of the population in the United States has emission levels comparable with the European middle 40%, despite being almost twice as poor as this group in purchasing power parity terms." Chancel, "Global Carbon Inequality over 1990–2019," 932.

77 Tim Gore, "Carbon Inequality in 2030: Per Capita Consumption Emissions and the 1.5°C Goal," *Oxfam* and *Institute for European Environmental Policy*, November 5, 2021. Available at https://www.oxfam.org/en/research/carbon-inequality-2030, joint policy paper by the two agencies.

78 Starr et al., "Income-Based U.S. Household Carbon Footprints (1990–2019)."

79 Ashfaq Khalfan, Astrid Nilsson Lewis, Carlos Aguilar, Jacqueline Persson, Max Lawson, Nafkote Dabi, Safa Jayoussi, Sunil Acharya et al., "Climate Equality: A Planet for the 99%," Oxfam, November 2023. Report available at https://policy-practice.oxfam.org/resources/climate-equality-a-planet-for-the-99-621551/

80 Beatriz Barros and Richard Wilk, "The Outsized Carbon Footprints of the Super-Rich," *Sustainability: Science, Practice and Policy* 17, no. 1 (2021): 316–322.

81 Tuvalu's emission data available at "Annual Greenhouse Gas Emissions Including Land Use," *Our World in Data*, www.ourworldindata.org; Lucy Craymer, "Tuvalu Turns to the Metaverse as Rising Seas Threaten Existence," Reuters, November 15, 2022, www.reuters.com.

82 Clint Wallace and Shelley Welton, "The Case for Taxing Luxury Emissions," *Regulatory Review* (October 28, 2024). www.theregreview.org.

83 Ilona M. Otto, Kyoung Mi Kim, Nika Dubrovsky, and Wolfgang Lucht, "Shift the Focus from the Super-Poor to the Super-Rich," *Nature Climate Change* 9 (2019): 82.

84 Dominik Wiedenhofer, Dabo Guan, Zhu Liu, Jing Meng, Ning Zhang, and Yi-Ming Wei, "Unequal Household Carbon Footprints in China," *Nature Climate Change* 7 (2017): 75–80.

85 Romain Sacchi, Viola Becattini, Paolo Gabrielli, Brian Cox, Alois Dirnaichner, Christian Bauer, and Marco Mazzotti, "How to Make Climate-Neutral Aviation Fly," *Nature Communications* 14 (2023): 3989; D. S. Lee, D. W. Fahey, A. Skowron, M. R. Allen, U. Burkhardt, Q. Chen, S. J. Doherty et al., "The Contribution of Global Aviation to Anthropogenic Climate Forcing for 2000 to 2018," *Atmospheric Environment* 244 (2021): 117834.

86 Stefan Gössling and Andreas Humpe, "The Global Scale, Distribution and Growth of Aviation: Implications for Climate Change," *Global Environmental Change* 65 (2020): 102194.

87 Yannick Oswald, Anne Owen, and Julia K. Steinberger, "Large Inequality in International and Intranational Energy Footprints Between Income Groups and Across Consumption Categories," *Nature Energy* 5 (2020): 231–239.

88 Gössling and Humpe, "The Global Scale, Distribution and Growth of Aviation."

89 Xinyi Sola Zheng and Dan Rutherford, "Aviation Climate Finance Using a Global Frequent Flying Levy." White Paper, The International Council on Clean Transportation, September 2022. https://theicct.org.

90 Martin Ravallion, Mark Heil, and Jyotsna Jalan, "Carbon Emissions and Income Inequality," *Oxford Economic Papers* 52, no. 4 (2000): 651–669.

91 Ravallion, Heil, and Jalan, "Carbon Emissions and Income Inequality," 667.

92 Michael D. Briscoe, Jennifer E. Givens, and Madeleine Alder, "Intersectional Indicators: A Race and Sex-Specific Analysis of the Carbon Intensity of Well-Being in the United States, 1998–2009," *Social Indicators Research* 155 (2021): 97–116; Nicole Grunewald, Stephan Klasen, Inmaculada Martínez-Zarzoso, and Chris Muris, "The Trade-off Between Income Inequality and Carbon Dioxide Emissions," *Ecological Economics* 142 (2017): 249–256; Andrew K. Jorgenson, Juliet B. Schor, Kyle W. Knight, and Xiaorui Huang, "Domestic Inequality and Carbon Emissions in Comparative Perspective," *Sociological Forum* 31, n. S1 (2016): 770–786; Andrew K. Jorgenson, Juliet B. Schor, Xiaorui Huang, and Jared Fitzgerald, "Income Inequality and Residential Carbon Emissions in the United States: A Preliminary Analysis," *Human Ecology Review* 22, no. 1 (2015): 93–106.

93 Andrew K. Jorgenson, Juliet Schor, and Xiaorui Huang, "Income Inequality and Carbon Emissions in the United States: A State-level Analysis, 1997–2012," *Ecological Economics* 134 (2017): 40–48.

94 Kyle W. Knight, Juliet B. Schor, and Andrew K. Jorgenson, "Wealth Inequality and Carbon Emissions in High-Income Countries," *Social Currents* 4, no. 5 (2017): 403–412.

95 Julius Alexander McGee and Patrick Greiner, "Can Reducing Income Inequality Decouple Economic Growth and CO2 Emissions," *Socius: Sociological Research for a Dynamic World* 4 (2018): 1–11. Some of these studies have been criticized on methodological grounds, underscoring the ambiguous state of the literature on inequality and greenhouse gas emissions. See Sebastian Mader, "The Nexus Between Social Inequality and CO2 Emissions Revisited: Challenging Its Empirical Validity," *Environmental Science and Policy* 89 (2018): 322–329. To complicate matters further, another study found that per capita emissions are positively associated with income inequality in developing countries, but not significantly associated in developed countries. Jiandong Chen, Qin Xian, Jixian Zhou, and Ding Li, "Impact of Income Inequality on CO2 Emissions in G20 Countries," *Journal of Environmental Management* 271 (2020): 110987.

96 James K. Boyce, "Inequality and Environmental Protection." Working Paper No. 52, Amherst: Political Economy Research Institute, University of Massachusetts, 2003; James K. Boyce, "Inequality as a Cause of Environmental Degradation," *Ecological Economics* 11, no. 3 (1994): 169–

178; James K. Boyce, "Is Inequality Bad for the Environment?," *Research in Social Problems and Public Policy* 15 (2008): 267–288.

97 Elinor Ostrom, "Frameworks and Theories of Environmental Change," *Global Environmental Change* 18, no. 2 (2008): 249–252; Laura Cushing, Rachel Morello-Frosch, Madeline Wander, and Manuel Pastor, "The Haves, the Have-Nots, and the Health of Everyone: The Relationship Between Social Inequality and Environmental Quality," *Annual Review of Public Health* 36 (2015): 193–209.

98 Francesco Vona and Fabrizio Patriarca, "Income Inequality and the Development of Environmental Technologies," *Ecological Economics* 70, no. 11 (2011): 2201–2213.

99 Jorgenson et al., "Income Inequality and Residential Carbon Emissions in the United States: A Preliminary Analysis."

100 Fergus Green and Noel Healy, "How Inequality Fuels Climate Change: The Climate Case for a Green New Deal," *One Earth* 5, no. 6 (2022): 635.

101 Green and Healy, "How Inequality Fuels Climate Change," 645.

102 Dario Kenner, *Carbon Inequality: The Role of the Richest in Climate Change* (Milton Park: Routledge, 2019); Otto et al., "Shift the Focus from the Super-Poor to the Super-Rich."

103 Manfred Lenzen, Ya-Yen Sun, Futu Faturay, Yuan-Peng Ting, Arne Geschke, and Arunima Malik, "The Carbon Footprint of Global Tourism," *National Climate Change* 8 (2018): 522–528.

104 Zan Yang, Shuping Wu, and Hiu Ying Cheung, "From Income and Housing Wealth Inequalities to Emissions Inequality: Carbon Emissions of Households in China," *Journal of Housing and the Built Environment* 32 (2017): 231–252.

105 Oswald, Owen, and Steinberger, "Large Inequality in International and Intranational Energy Footprints."

106 Wallace and Welton, "The Case for Taxing Luxury Emissions"; Grunewald et al., "The Trade-off Between Income Inequality and Carbon Dioxide Emissions"; Jorgenson, Schor, and Huang, "Income Inequality and Carbon Emissions in the United States: A State-level Analysis, 1997–2012."

107 Grunewald et al., "The Trade-off Between Income Inequality and Carbon Dioxide Emissions," 253.

108 Likewise, one could also use the proceeds from progressive taxes to support targeted programs to improve the lives of poor people—such as improving public transportation or providing energy efficient housing—rather than merely offering lump sum transfers. Zachary Liscow and Abigail Pershing, "Why Is So Much Redistribution In-Kind and Not in Cash? Evidence from a Survey Experiment," *National Tax Journal* 75, no. 2 (2022): 313–354.

109 Milena Büchs, Noel Cass, Caroline Mullen, Karen Lucas, and Diana Ivanova, "Emissions Savings from Equitable Energy Demand Reduction," *Nature Energy* 8 (2023): 758–769.
110 Zachary Liscow, "Is Efficiency Biased?," *University of Chicago Law Review* 85, no. 7 (2018): 1649–1718.
111 Fergus and Healy, "How Inequality Fuels Climate Change."
112 Laura A. McKinney and Gregory M. Fulkerson, "Gender Equality and Climate Justice: A Cross-National Analysis," *Social Justice Research* 28, no. 3 (2015): 293–317; Christina Ergas and Richard York, "Women's Status and Carbon Dioxide Emissions: A Quantitative Cross-National Analysis," *Social Science Research* 41, no. 4 (2012): 965–976.
113 Astghik Mavisakalyan and Yashar Tarverdi, "Gender and Climate Change: Do Female Parliamentarians Make Difference?," *European Journal of Political Economy* 56 (2019): 151–164.
114 Christina Ergas, Patrick Trent Greiner, Julius Alexander McGee, and Matthew Thomas Clement, "Does Gender Climate Influence Climate Change? The Multidimensionality of Gender Equality and Its Countervailing Effects on the Carbon Intensity of Well-Being," *Sustainability* 13 (2021): 3956.
115 Zhike Lv and Chao Deng, "Does Women's Political Empowerment Matter for Improving The Environment? A Heterogeneous Dynamic Panel Analysis," *Sustainable Development* 27, no. 4 (2019): 603–612.
116 Zeynep Pamuk, "Climate Change, Inequality, and Expert Knowledge," in this volume.
117 Arthur M. Okun, *Equality and Efficiency: The Big Tradeoff* (Washington, DC: Brookings Institution, 1975), 91.
118 Duncan Kennedy, "Law and Economics from the Perspective of Critical Legal Studies," in *The New Palgrave Dictionary of Economics and the Law*, ed. Peter Newman (New York: Macmillan Reference Limited, 1998), 465–474.
119 Alyssa Battistoni, "Domination in the Age of the Externality," in this volume.
120 Michael Greenstone, Christian Leuz, and Patricia Breuer, "Mandatory Disclosure Would Reveal Corporate Carbon Damages," *Science* 381 (2023): 837–840.
121 Robert M. Solow, *An Almost Practical Step Toward Sustainability* (New York: Resources for the Future, 1992), 8–9.
122 Bernardo A. Bastien-Olvera and Frances C. Moore, "Use and Non-Use Value of Nature And the Social Cost of Carbon," *Nature Sustainability* 4 (2021): 101–108.
123 Steve Keen, "The Appallingly Bad Neoclassical Economics of Climate Change," *Globalizations* 18, no. 7 (2020): 1149–1177, at 1151–1153.

Even if the claim that climate change will only impact outdoor industries were somehow accurate, implicit in the approach is an assumption that, because agriculture only contributes around 4% of global GDP, we can get along fine without it.

124 Herman E. Daly, *Ecological Economics and the Ecology of Economics* (Cheltenham: Edward Elgar Publishing, 1999), 50; Douglas A. Kysar, "Climate Change, Cultural Transformation, and Comprehensive Rationality," *Boston College Environmental Affairs Law Review* 31, no. 3 (2004): 555–590.

125 Robert Costanza, John H. Cumberland, Herman Daly, Robert Goodland, and Richard B. Norgaard, *An Introduction to Ecological Economics* (Boca Raton, FL: CRC Press LLC, 1997), 80.

126 Robert Fletcher and Crelis Rammelt, "Decoupling: A Key Fantasy of the Post-2015 Sustainable Development Agenda," *Globalizations* 14, no. 3 (2016): 450–467.

127 Douglas A. Kysar, "Law, Environment, and Vision," *Northwestern University Law Review* 97, no. 2 (2004): 675–730; Douglas A. Kysar, "Sustainability, Distribution, and the Macroeconomic Analysis of Law," *Boston College Law Review* 43, no. 1 (2001): 1–71.

128 Kate Raworth, *Doughnut Economics: Seven Ways to Think Like a Twenty-First Century Economist* (White River Junction, VT: Chelsea Green Publishing).

129 Daly, *Ecological Economics and the Ecology of Economics*, 4.

130 Pamuk, "Climate Change, Inequality, and Expert Knowledge."

131 Helmut Haberl, Dominik Wiedenhofer, Doris Virág, Gerald Kalt, Barbara Plank, Paul Brockway, Tomer Fishman et al., "A Systematic Review of the Evidence on Decoupling of GDP, Resource Use and GHG Emissions, Part II: Synthesizing the Insights," *Environmental Research Letters* 15, no. 6 (2020): 065003; Fletcher and Rammelt, "Decoupling: A Key Fantasy of the Post-2015 Sustainable Development Agenda."

132 Jefim Vogel and Jason Hickel, "Is Green Growth Happening? An Empirical Analysis of Achieved Versus Paris-Compliant CO_2–GDP Decoupling in High-Income Countries," *Lancet Planetary Health* 7 (2023): e759–e769.

133 Tribe, "Ways Not To Think About Plastic Trees," 1324.

134 "What has been omitted is, at base, an appreciation of an ancient and inescapable paradox: We can be truly free to pursue our ends only if we act out of obligation, the seeming antithesis of freedom. To be free is not simply to follow our ever-changing wants wherever they might lead. To be free is to choose what we shall want, what we shall value, and therefore what we shall be." Tribe, "Ways Not To Think About Plastic Trees," 1326–1327.

135 William D. Nordhaus, "Climate Change: The Ultimate Challenge for Economists," *American Economic Review* 109, no. 6 (2019): 1991–2014.

136 Tim Wu, "Did Laurence Tribe Sell Out?," *New York Magazine* (May 6, 2015).

137 Christina Pazzanese, "Laurence Tribe Speaks on His Career in Constitutional Law," *Harvard Gazette* (June 24, 2020). https://news.harvard.edu.

138 Andrew Rice, "How the President's Longtime Confidant Became His Greatest Adversary on Climate Change," *New York Magazine* (July 28, 2015). https://nymag.com.

139 William Boyd, "Ways of Seeing in Environmental Law: How Deforestation Became an Object of Climate Governance," *Ecology Law Quarterly* 37, no. 3 (2010): 843–916.

2

CLIMATE CHANGE, INEQUALITY, AND EXPERT KNOWLEDGE

ZEYNEP PAMUK

The neoliberal paradigm in climate policy, built on principles of welfare economics and cost-benefit analysis, has been criticized for its narrow focus on individual self-interest and preference satisfaction, taking existing property rights, income distributions, population sizes, adaptation capacities, and technological pathways as a given. For decades, policymakers continued to tinker with existing economic models, allowing incremental moves from the status quo but precluding radical change. Given that our current pathway might lead to a global disaster in the near future, critics have pointed out that it is high time that we look beyond narrow economic questions such as the correct social cost of carbon or discount rate and expand our political imagination. One of the most interesting alternatives on offer involves reimagining inequality reduction as a form of climate mitigation.[1] On this view, we should stop thinking of efficiency and equity as separate dimensions of climate policy, but rather think of the latter as a way of achieving the former. Limiting the emissions of the rich and redistributing resources through progressive taxation should be viewed as forms of climate policy.

In this chapter, I will first distinguish between two critiques of the cost-benefit paradigm that are not sufficiently distinct in this debate: one that criticizes the narrow focus and assumptions of welfare economics and another that more broadly questions the role of experts in shaping and limiting our political imagination. Despite its novelty and promise, I will argue that the proposal of reducing inequality as a form of climate mitigation is not immune from the latter critique. What we know about the relationship between inequality and emissions comes from expert models, and the experts disagree on whether reducing inequality will prove an effective climate mitigation strategy and, if so, through which economic and social policies. Proponents of three dominant approaches to climate policy—green growth, a Green New Deal, and degrowth—make different assumptions about key unknowns, each influenced by their own economic and political views.

I then address the question of how democracies can take action on climate change in the face of ongoing expert disagreement. Contra those who have called for experts to set aside their disagreements for the sake of stopping environmental catastrophe,[2] I maintain that scientific disagreements should be highlighted in public debates around climate change and scrutinized through institutions designed to allow the public and policymakers to understand the strengths and limits of expert views. We should seek to both normalize and institutionalize scientific disagreement in the public sphere and allow the public and policymakers to make up their own mind among rival proposals. Institutions that allow democratic scrutiny of science would increase the accountability of experts as well as checking their influence over our political imagination. Finally, since expert decisions made earlier in the research process set the bounds of what we consider to be possible and desirable, I argue that we must also increase democratic input into science at earlier stages of the research process.

Inequality Reduction as Climate Policy

I want to begin by distinguishing between two critiques of the cost-benefit paradigm, which are often conflated. The first critique squarely targets welfare economics—what it takes for granted, how it conceptualizes human beings and their relationship to the environment, what it counts as a cost and what a benefit, and how it devalues things it cannot measure, taking growth as an imperative and valorizing efficiency above all else. These call for a new policy paradigm and a new approach to economics. The question is which strategies can effectively get us there. The second critique is less clearly articulated but further reaching: It targets expert-driven approaches to policymaking as a whole. Economists and policymakers are criticized for burying their values behind modeling assumptions, thus removing them from political contestation.[3] The problem is not just the content of the assumptions—different administrations plug different numbers into the models and thereby generate vastly different policies. It is also that this covert politics is conducted by experts within the administrative state, based on models devised by social scientists, shielded from proper public scrutiny and debate. The models and findings of experts drive the conversation on climate change, and the limits of their studies end up being the limits of our collective political imagination.

New ideas for moving beyond the neoliberal paradigm are themselves often susceptible to this critique. Kysar's proposal to view inequality reduction as climate mitigation is one example.[4] The proposal rests on a series of facts that illustrate the severe inequalities in emissions not only between rich and poor countries but between the rich and poor within each country. Kysar cites research that suggests that inequality itself—rather than just absolute levels of income and wealth—could be a driver of higher emissions.[5] He then argues that reducing inequality through limits on luxury emissions and steeply progressive taxation could be a form of climate change policy, rather than merely a policy of redistribution.

However, there is some dispute in the literature about how broadly the relationship between inequality and emissions holds, and we depend on conflicting social scientific studies to determine the magnitude and scope of the association. Two early studies on the topic found a negative association between reducing inequality and reducing greenhouse gases, suggesting that there is actually a tradeoff between climate mitigation and inequality.[6] A few studies found no significant relationship at all.[7] The extent to which findings of a positive relationship between inequality and emissions apply beyond high-income countries is also unclear.[8] Since reducing inequality in middle- and lower-income countries tends to increase total consumption, it might increase greenhouse gas emissions as well. To make the point actionable, we need more empirical research that establishes the relationship conclusively. But setting aside the difficulties of obtaining the right evidence, the real challenge only starts with the observation that there is an association between inequality and emissions. Whether reducing inequality will in fact prove an effective climate mitigation strategy depends on which economic and social policies will be adopted to reduce inequality and how these interact with other policies that are or could be adopted to reduce greenhouse gases.

The literature is divided on these points. Three approaches currently dominate the discourse—green growth, Green New Deal, and degrowth—and there is serious disagreement among their advocates. Green growth appears likely to increase inequality by increasing unemployment and the relative demand for high-skilled workers; it does not prioritize inequality reduction.[9] But it might turn out to do just as well if not better than more egalitarian policies in terms of emissions reductions, thus potentially revealing a tradeoff between low emissions and low inequality. If we want to focus on inequality reducing policies, the appropriate comparison is between the Green New Deal—which is essentially a more egalitarian green growth—and degrowth. One socialist economist argues that degrowth would create soaring poverty and unemployment, with only modest emissions reductions.[10] By contrast, a recent paper by four ecological economists in *Nature* argues that degrowth will deliver the greatest gains both in terms of reducing inequality and in reducing greenhouse gas emissions—the only thing we would have to give up is growth.[11]

The disagreement is hard to resolve because each camp makes different assumptions about key unknowns. For instance, some critics assume that a Green New Deal is likely to run into problems from the increased need for mining for minerals such as copper, nickel, cobalt, and lithium, which they claim will result in both more inequality and more emissions, as well as other forms of ecological damage.[12] Defenders of these proposals also disagree on rates of future technological change, for instance on issues such as how quickly electric vehicle batteries will improve and cities will be able to clean their electrical grids, and whether technological efficiency gains will have rebound effects that end up increasing emissions—a phenomenon known as the Jevons paradox. More foundational disagreements include questions such as whether capitalist growth is compatible with sustainability,[13] and whether large capitalist societies can transition to a steady state without crises of poverty and unemployment.[14] What we know about the relationship between emissions reduction and inequality—as well as their relationship to growth, unemployment, ecosystem health and other things we care about—depends on models with highly speculative and contradictory assumptions. While the terms of this conversation are surely an improvement over the neoliberal one as far as climate policy is concerned, the dependence on experts is still a constraint on properly democratic discussion and action.

My aim is not to give a comprehensive review of this debate but to draw attention to the role that disagreeing experts play in shaping political deliberation and the resulting deadlock even among those committed to significant action on climate change. A 2022 editorial in *Nature* warned that the rift among research communities had become an impediment to action and called on researchers to set aside their disagreements to stop environmental destruction.[15] This view assumes that scientific disagreement hinders policymaking and that consensus among scientists is a precondition to political action. Scientists are expected to resolve disputes among themselves and present a united front to policymakers, who can then make policy on the basis of "the science." The authority and usefulness of science in the public sphere are taken to rest on the ability of science to present a consensus view to the public and policymakers.

This popular model of the relationship between scientific disagreement and political action is problematic for a few reasons. First, forcing scientists to resolve or somehow set aside deep disagreements before communicating with policy makers creates an epistemic loss. The consensus arrived at in this process will be unlike a scientific consensus in which different studies converge upon the same results through different evidence. Disagreements over background assumptions and the interpretation of findings reveal the uncertainty and limits of each view. Differences between proponents of green growth, a Green New Deal, and degrowth contain information about possible economic, technological, and ecological futures, as well as pointing to the gaps in our current knowledge. Suppressing these deprives policymakers of valuable information. Secondly, this approach delegates essentially political judgments to experts. The assumptions that go into these models, as well as their definition of core concepts, such as sustainability or growth, are value laden. Not only do different models imply different political programs, but their assumptions and concepts are also rooted in different viewpoints. This is particularly pronounced in scientific models that involve complex human systems such as the economy. These scientific disputes cannot be resolved without settling disagreements over values, and disagreements over values are best resolved through political processes. Scientists' superior knowledge does not qualify them to make these judgments in others' name.

Climate Policy and Expert Knowledge

How, then, can democracies deal with expert disagreement over climate policies? Instead of asking researchers to set aside disagreements for the benefit of policymakers, my proposal is that scientific disagreements should be oriented toward a public audience, through institutions designed to allow the public and policymakers to scrutinize expert findings and understand their strengths and limits. The aim should be to normalize and institutionalize scientific disagreement in the public sphere and allow the public and policymakers to make up their own minds. Revealing the values and assumptions that shape the research would inform democratic debate and serve as an organizing informational base for partisan politics. Institutions that allow democratic scrutiny of science would increase experts' accountability and check their influence in shaping and constraining political possibilities.

One example of such an institution is an adversarial "science court."[16] Proponents of green growth, degrowth, and a Green New Deal could debate one another in front of a jury of randomly selected citizens, exposing the sources of disagreement between them and the limits and uncertainty of each view. The jury would deliberate and choose between the proposals. The court would not be intended to settle the facts for the scientific community but to make a policy recommendation, evaluating the scientific evidence alongside the values. Such a process would allow citizens to assess the claims of experts while bringing their own values and perspectives to the deliberation. It would also make citizens feel less removed from an area of policy currently dominated by experts.

I anticipate the concern that institutions highlighting scientific disagreements in the public sphere might reduce public trust in scientists. This is particularly salient in the climate change case since the attention paid to denialists distorted public perceptions of climate science and eroded the credibility of the field for a long time. It is important, however, to distinguish the creation of a false equivalence between a majority of studies showing the existence of anthropogenic climate change and a small number denying it, from issues of genuine division within the scientific community. In the latter case, drawing attention to points of disagreement will allow trust to be placed where it is merited. The aim of political institutions that deal with expertise should not be to encourage unthinking deference to experts but to cultivate a healthy skepticism. Carefully designed institutions that channel and redirect an attitude of skeptical questioning can facilitate judgment about where trust is appropriately placed and legitimate expert influence in the political process.

While democratic scrutiny of expert advice can allow the public and policymakers to make their own judgments, expert decisions made earlier in the research process determine which knowledge becomes available and thus draw the limits of our political imagination. What we consider to be possible or desirable is shaped by the research scientists choose to undertake. We can't intuit the relationship between inequality, carbon emissions, and growth; we need evidence. Before social scientists began to pay attention to this relationship in the 2000s, inequalities within countries were largely absent from the conversation on climate change, despite the long-standing emphasis on global inequalities in emissions. The relationship between gender inequality and climate change is another example. Conceiving of gender inequality as climate mitigation became possible only once studies showed that improving the status of women reduces emissions.[17]

How can democratic societies address the way political imagination is shaped and constrained by choices made earlier in the research process? One response is to argue that certain policies are good or bad, ethical or unethical, regardless of their consequences. We don't need expert predictions about their effects in order to evaluate them. For instance, we might categorically rule out exporting waste to developing countries because it is morally wrong under all circumstances. Versions of the degrowth case veer into categorical moral arguments against consumption or denouncements of capitalist growth as inherently incompatible with an ethical and harmonious relationship to nature.[18] This approach can be attractive to some, but its reach is limited. To win over people who are not persuaded by the categorical case, it helps to have arguments about the expected consequences of policies. In general, most policy disputes cannot be settled on a priori grounds. We need not take a narrowly economic view on which variables matter, but we still need to know what consequences to expect and what tradeoffs we will face. It is difficult to avoid dependence on experts for these questions.

Another solution is to find experts who possess a wider imagination. This draws attention to the training and selection of academics. We want people with different backgrounds, experiences, values, and viewpoints to become scientists and social scientists. Uniformity of perspectives and backgrounds is a serious threat to academic creativity, which in turn limits political possibilities. If everyone in a professional field holds the same values, it will be difficult to challenge scientific paradigms from within. This might be one explanation for the persistence of the neoliberal framework. The move from neoliberal economics to socialist climate proposals shows that such a change may already be at work, though many of these challenges have come from outside mainstream economics departments.

The limitation of this approach is that academia trains its members to think differently than non-experts and converts newcomers to dominant paradigms, whether neoliberal or socialist or something else. This is professionalization. It has benefits in terms of research productivity and advancing paradigms, but also creates a gap between academics and laypeople. We can't just expect that a demographically diverse group of experts will generate a wide enough range of ideas. Members of the public should also have a say in determining the direction of scientific research. This requires paying more attention to the distribution of science funding and involving those outside professional fields in these processes.

If democratizing agenda setting is part of the challenge, the other is opening up the assumptions of economic models to public deliberation at earlier stages of the research process. There must be more participatory decision-making on issues such as whether we want radical or conservative assumptions in models, how much behavioral change and individual responsibility we should assume, and how much we want to avert certain bad outcomes. Broader engagement on these decisions could allow us to move away from a policy model of warring experts pushing their own ideological commitments through modeling assumptions, toward one where scholars learn from and apply the perspectives of lay communities, while submitting their own values to public scrutiny.

The degrowth movement offers a promising example since its animating ideas were developed and influenced by a combination of activists, grassroots movements, and academics, and it has migrated into more mainstream academic models and policy conversations. Degrowth conferences bring together scientists, policymakers, and activists in participatory formats and tend to acknowledge the importance of the input of local communities in producing and questioning knowledge.[19] Regardless of what we conclude about the merits of the proposal, its mode of knowledge production can be a model for other expert-layperson interactions.

NOTES

1 Douglas A. Kysar, "Ways Not to Think About Climate Change," this volume.

2 "Are There Limits to Economic Growth? It's Time to Call Time on a 50-Year Argument," *Nature* 603 (2022): 361.

3 Kysar, "Ways Not to Think About Climate Change."

4 Kysar, "Ways Not to Think About Climate Change."

5 Kysar, "Ways Not to Think About Climate Change."

6 Nico Heerink, Abay Mulatu, and Erwin Bulte, "Income Inequality and the Environment: Aggregation Bias in Environmental Kuznets Curves," *Ecological Economics* 38, no. 3 (2001): 359–367; Martin Ravallion, Mark Heil, and Jyotsna Jalan, "Carbon Emissions and Income Inequality," *Oxford Economic Papers* 52, no. 4 (2000): 651–669.

7 Simone Borghesi, "Income Inequality and the Environmental Kuznets Curve," in *Environment, Inequality, and Collective Action*, ed. Basili Marcello, Franzini Maurizio, and Vercelli Alessandro (New York: Routledge, 2006), 33–51; Martin Gassebner, Michael J. Lamla, and Jan-Egbert Sturm, "Determinants of Pollution: What Do We Really Know?," *Oxford Economic Papers* 63, no. 3 (2011): 568–595.

8 Andrew K. Jorgenson, Juliet B. Schor, Kyle W. Knight, and Xiaorui Huang, "Domestic Inequality and Carbon Emissions in Comparative Perspective," *Sociological Forum* 31, no. S1 (2016): 770–786.

9 Simone D'Alessandro, André Cieplinski, Tiziano Distefano, and Kristofer Dittmer, "Feasible Alternatives to Green Growth," *Nature Sustainability* 3, no. 4 (2020): 329–335.

10 Robert Pollin, "De-Growth vs a Green New Deal," *New Left Review* 112 (2018): 5–25.

11 D'Alessandro et al., "Feasible Alternatives to Green Growth."

12 Bill McKibben, "To Save the Planet, Should We Really Be Moving Slower?," *The New Yorker* (July 5, 2023). https://newyorker.com.

13 Tim Jackson and Peter A. Victor, "Unraveling the Claims for (and against) Green Growth," *Science* 366, no. 6468 (2019): 950–951.

14 Pollin, "De-Growth vs a Green New Deal."

15 "Are There Limits to Economic Growth?"

16 Zeynep Pamuk, *Politics and Expertise: How to Use Science in a Democratic Society* (Princeton, NJ: Princeton University Press, 2021).

17 See, e.g., Astghik Mavisakalyan and Yashar Tarverdi, "Gender and Climate Change: Do Female Parliamentarians Make Difference?," *European Journal of Political Economy* 56 (2019): 151–164.

18 Giorgos Kallis, *Limits: Why Malthus Was Wrong and Why Environmentalists Should Care* (Stanford, CA: Stanford University Press, 2019).

19 Giorgos Kallis, Vasilis Kostakis, Steffen Lange, Barbara Muraca, Susan Paulson, and Matthias Schmelzer, "Research on Degrowth," *Annual Review of Environment and Resources* 43 (2018): 291–316.

PART II

CLIMATE CHANGE AND THE CAPITALIST ORDER

3

DOMINATION IN THE AGE OF THE EXTERNALITY

ALYSSA BATTISTONI

While as recently as a decade ago climate change was still viewed as a problem of the distant future, today it is one whose effects are plainly obvious: in punishing heat waves that have swept through India and Europe; wildfires in Australia and smoke clogging North American cities; megadroughts in the Horn of Africa and flooding in Pakistan; cyclones striking Mozambique and hurricanes devastating Caribbean island states. Thus far temperatures have risen (only) 1.1°C from preindustrial temperatures; they are projected to rise 2.5°C or more by the end of the century. In political discourse it has become routine in recent years, to the point of cliché, to describe climate change as an "existential threat." Within economics, however, these extraordinary transformations are frequently traced to a technical and seemingly minor cause: the absence of price.[1] As the economist Nicholas Stern famously observed in 2007, "greenhouse gas emissions are externalities, and represent the biggest market failure the world has seen."[2] Externalities occur when economic activity causes costs for third parties that are not reflected in the costs to the producer, such that they are not taken into account in decision-making. The solution to climate change, on this view, consists in "internalizing the externality": incorporating unpaid costs into prices so that the market can work as promised. The externality has been at the heart of major environmental policy frameworks since the late twentieth century, and climate policy especially, most prominently via carbon taxes and cap-and-trade programs.[3] These have, in turn, been taken up in moral philosophy oriented toward policy recommendations and decried by critics of economism.[4] Yet the concept of the externality itself has gone strikingly unexamined in political theory and philosophy.[5]

This chapter argues that the externality offers a rich entry point to a political economic analysis of climate change, one that casts debates about environmental injustice and moral responsibility in a new light. The externality, after all, is fundamentally concerned with the divergence between private and social interests; between individual and collective action; and between intentions and consequences—all themes of vital interest to current debates about climate change, as well as to thinking about politics writ large. It opens up major questions about how intent, agency, and responsibility are organized by the political economic system upon which nearly all human beings now rely for our livelihoods—capitalism, a system that now stands alone.

Indeed, attention to the externality is only the latest iteration of a long-standing debate among theorists of political economy about the significance of unintended consequences and the relationship between the political and economic spheres.[6] In the eighteenth century, Adam Smith famously argued that the pursuit of self-interest tended, however unintentionally, to generate collective wellbeing, whereas intentional state action frequently undermined it; in the nineteenth century, Karl Marx "stood Adam Smith on his head" by pointing to the perverse unintended consequences of self-interested action in the market.[7] Capitalism's recurring crises, for Marx, were just as much the unintended product of aggregated market choices as its generation of wealth; a socialist society, by contrast, would govern production more rationally. By the mid-twentieth century, anxiety about unintended consequences had become pervasive: variously described in terms of counterfinality, tragedy, and reflexive modernity.[8] Notably, many of these theories, even when articulated by non-environmental thinkers, deal explicitly with the effects of action on the natural world, illustrated by examples of peasants, shepherds, pollution, or fishermen. It should not surprise us, then, that on closer inspection, the history of the externality in economic thought is at its heart the history of economists encountering the environment: in other words, of economists seeking to grapple with and contain the effects of economic activity in the material world amidst shifting political and economic circumstances. Similarly, it is no coincidence that the history of the concept is effectively contemporaneous with that of the Great Acceleration: Although first raised in the early twentieth century as a minor flaw to be redressed by economists, the externality would only become the subject of sustained attention in the face of explosive postwar growth, before coming, in the early twenty-first century, to particular prominence in debates about climate change.[9]

Upon stepping back from the epistemic framework generated by this contingent history, however, it is remarkable that a phenomenon described as an "existential threat" to humanity has been reduced to a problem of missing prices, as Douglas Kysar also observes.[10] It is only under conditions in which most things are bought and sold that this could be the case—and this is a remarkably recent feature of human history. Most theorists of the externality treat markets as an ideal type of allocation mechanism—a means by which goods (or bads) might be distributed via exchanges negotiated among individual actors. But the condition of generalized market dependence, in which most people work for wages and obtain most of what they need to survive through exchange rather than through subsistence activity, is a unique and defining feature of capitalism in particular as a system of political and economic organization.[11] It is this dependence that makes the prospect of market failure so threatening—and so rich for political interrogation. In systems of logic or infrastructure, it is often the points of failure that are most revealing, and this is no less true of so-called market failure. Although externalities are frequently treated as an exception to the rule, they illuminate the rules themselves: how markets are *supposed* to work. The externality, in other words, is not an error or absence in the market, but rather an extreme example of how markets *normally* function, and what they are supposed to do. In looking more closely at the externality, this chapter reframes two widely discussed features of climate politics: the uneven distribution of negative effects and the challenges of assessing "fragmented" moral agency and "distributed" responsibility.

Within political philosophy, climate change is perhaps most widely understood as a problem of justice in distribution. It is frequently noted that both the benefits of carbon-intensive activities and harms resulting from the growing concentration of atmospheric carbon and planetary warming are unevenly distributed, in ways that tend to exacerbate existing forms of inequality and oppression. Carbon emissions are highly correlated with wealth, while the effects of climate change are likely to impact the poor (who have emitted least) first and hardest.[12] Many have argued that high carbon emitters have a moral obligation to reduce emissions, at either the individual or collective level, while often calling for a reevaluation of agency itself in light of what climate change seems to reveal about its "fragmented" or "distributed" quality.[13] In these discussions, climate change is frequently understood, following Derek Parfit, as an "aggregation problem"—the unexpected result of a huge number of individually harmless decisions.[14] By treating environmental problems primarily as a matter of scale (and thus, effectively, as matters of population), however, these accounts tend to draw too direct a link between action and effect, one that fails to address the way that individual agency is channeled and mediated by both social institutions and the material world.

I attend instead to what Iris Marion Young describes as the "social structural processes" that shape our relationships to one another—and that inform our relationships to the more-than-human world.[15] Read thus, the externality points not only to the uneven distribution of "costs and benefits" but to the foundational inequality in social power that is constitutive of capitalist societies, and the structuring force of the market. Instead of understanding the resulting harms through the familiar frames of distributive justice and individual agency, I make the case for seeing climate change in terms of *domination*. Domination is, admittedly, not an intuitively applicable concept. In the republican tradition, it is typically understood as the arbitrary power of one person over another, considered in terms of intentional, interpersonal action.[16] Unintentional effects of action, structural forces, and natural elements are not, in republican accounts, typically understood as sources of domination. While neo-republican theorists have productively extended the classical account of domination to analyze power in the workplace and class hierarchies, they have struggled to come to terms with other, more diffuse constraints on action. While my account draws on and adds to recent neo-republican work in political theory, then, it also offers a novel analysis.[17]

My account of domination in this chapter therefore contains two dimensions. The first builds on neo-republican theories of what

Alex Gourevitch describes as the "structural domination" of class society, showing how this form of domination permeates aspects of economic life beyond the workplace or labor-capital relationship. The second addresses the condition of what some theorists have called "abstract" or "social domination"—a form of unfreedom generated by market relations, and one that fits less easily with the traditional republican view.[18] While republicans have often read markets as counters to arbitrary power, I argue that the distinctive forms of constraint imposed by markets demand a novel analysis. I draw on Jean-Paul Sartre's concept of seriality and Eric MacGilvray's analysis of market freedom as "nonresponsibility" to describe a condition that actively undercuts our ability to make meaningful choices about our lives, both individually and collectively—what I here call "market domination."[19] As I argue in more detail elsewhere, both of these forms of domination are *materially mediated*: Human action is not only channeled through social institutions, as described here, but also expressed in and through the material world.[20]

In contrast to work in moral philosophy that brings ideal theory to bear on policy, this chapter does not offer a particular policy recommendation. Nor does it reject outright policies like carbon taxes and markets that propose to "internalize the externalities." Similar policies have worked to address other environmental problems understood via the externalities framework, as Madison Condon points out in the chapter herein with respect to sulfur dioxide. While I disagree with Condon's suggestion in her contribution to this volume that climate change does not actually constitute an externality, I share her skepticism that such measures will achieve decarbonization goals on their own, even as they might plausibly operate as part of a broader program for decarbonization, as Mark Budolfson's chapter suggests. Similarly, although the critique offered here locates the problem in structural features of capitalism, it does not thereby argue that capitalism must be entirely transcended in order to avert climate catastrophe—although I argue that it must be confronted more directly.[21] I agree, for example, with Kysar's suggestion that limiting wealth and inequality ought to be seen as climate mitigation measures.[22] But this chapter is pitched at a different level. It is an effort to identify the logics, conflicts, and limits of such policies, on the view that understanding systemic forces, and how policy proposals operate within them, is a necessary complement to—rather than substitute for—more pragmatic measures.

A Critical History of the Externality

A Pall over Liberalism: Arthur Pigou and the Birth of the Externality

Writing at the height of England's early and tumultuous industrialization, the British welfare economist Arthur C. Pigou (1877–1959) noted that the production of commodities was often accompanied by unintentional but sometimes severe material effects. To theorize this problem he built on a conceptual architecture inherited from the "marginalist revolution" of the 1870s, which had abandoned substance theories of value in favor of the subjective judgment of personal utility.[23] Early twentieth century welfare economics, to which Pigou was a pioneering contributor, sought to integrate these methodological insights with utilitarian ethics, with the goal of developing a truly scientific study of social welfare.[24] Prices were crucial to this project: They were, as the highly influential English marginalist Alfred Marshall (and Pigou's teacher) observed, "the one convenient means of measuring human motive on a large scale."[25] Marshall nevertheless freely admitted that money was a crude measure that failed to capture all effects of economic activity. This difficulty was recognized by many economists, but first discussed in systematic detail by Pigou's classic *The Economics of Welfare* (1920).[26] Following Marshall, Pigou argued that assessments of economic welfare had to use "the measuring rod of money," even if some things were beyond its scope.[27] He acknowledged, however, that this method sometimes produced "violent paradoxes" wherein welfare and price diverged.[28] He described such instances, where prices failed to reflect the effects of production on society at large, as "external economies."

The valence of the "external economy" was not always negative: Sometimes private producers accidentally generated unpriced social benefits.[29] Pigou's central example of an external effect, however—destined to become the textbook case—was a negative one: a factory with a smoky chimney.[30] The chimney smoke imposed costs on the community at large—"in injury to buildings and vegetables, expenses for washing clothes and cleaning rooms, expenses for the provision of extra artificial light, and in many other ways"—which were not reflected in the costs to the factory owner.[31] While thinkers like Adam Smith and Bernard Mandeville had famously proposed that the pursuit of individual self-interest was the best way of increasing the common good, Pigou argued that in certain cases the pursuit of private wealth tended to *diminish* public welfare rather than increasing it.[32] Fortunately, these problems seemed to be relatively rare and easily rectified. Where the market failed to secure social benefits, Pigou argued, the state was justified in intervening to address the disparity. It could estimate the costs of external effects and incorporate them into the price of relevant goods through a tax or similar pricing mechanism.[33]

For the next several decades most economists followed Pigou's view of externalities as an instance of "market failure" in which markets failed to optimally allocate resources, but a negligible one that could be solved with minor adjustments. Externalities remained a footnote to the canons of price theory in this period: They were, in the words of one midcentury welfare economist, "exceptional and unimportant."[34] As postwar economic growth and material throughput skyrocketed, however, pollution problems emerged or accelerated across the industrialized world.[35] Externalities suddenly began to appear ubiquitous and significant—and a concomitant economic literature exploded. In this context, Pigou's suggestion that externalities constituted a potentially systematic "market failure" began to seriously concern champions of free markets, and the emerging formation of neoliberal economists of the Chicago and Virginia Schools in particular. Pigou's account of disparities between private and public wellbeing seemed to cast a smoggy pall over the happy Mandevillian marriage of the individual and common good. Some economists went so far as to suggest that these "social costs" undermined the case for private enterprise altogether. This was not necessarily a problem for economics itself, which had largely rejected the idea that social welfare could be calculated in the aggregate and adopted in its place the far narrower standard of Pareto efficiency. But members of the public were often concerned with public welfare, in the broad sense, even if economists were not.[36]

The externality was also a problem for the increasingly hegemonic view of markets as expressions of individual liberty.[37] Markets were championed as sites of consensual exchange, in contrast to the coercive force of the state—and yet externalities imposed costs on people who had *not* consented to bear them. People forced to breathe particulate matter had not agreed to do so; nor were they compensated with a share of the benefits enjoyed by those generating pollution. Why was this nonconsensual infringement on bodily autonomy acceptable where, say, forced labor was not? The externality framework highlighted a proliferating number of cases in which exchange appeared to violate a liberal tenet as foundational as John Stuart Mill's harm principle.[38] No less a libertarian than Robert Nozick would struggle, in his *Anarchy, State, and Utopia* (1974), to reconcile a moral framework organized around the inviolable Kantian individual with the fact that nearly all actions have effects extending beyond parties to a contract, some of which might harm others.[39] Even Ludwig von Mises accepted the state's role in "guaranteeing the protection of life, health, liberty, and private property"; even Friedrich Hayek followed Pigou in granting the state a role in regulating the "smoke and noise of factories."[40] But if externalities were truly ubiquitous, they threatened to license a drastic extension of government and severe restrictions on market freedom. As complaints about the "smoke nuisance" intensified, Milton Friedman noted that "there is no transaction between individuals that does not affect third parties to some extent, however trivial, so there is literally no governmental intervention for which a case cannot be offered along these lines."[41] The externality, he worried, could therefore be "used to justify a completely unlimited extension of government."[42]

Bargaining Defended: Ronald Coase and the Social Cost

As attention to the externality problem had grown, however, so had scrutiny of Pigou's theory. In 1960, the British economist Ronald Coase launched a major, direct critique of Pigou in his landmark article "The Problem of Social Cost."[43] The paper would become a foundational text in the law and economics tradition, and among the most cited in all legal scholarship. Coase's critique drew on the "new" welfare economics: Pigou's utilitarianism, Coase argued, had led him to import a moral framework that informed his assessment of both the necessity and ends of state intervention. By stating as a matter of fact that certain private activities caused public injury—that when a factory's "smoky chimney" affected the surrounding air, for example, it constituted a clear case of social harm caused by the factory, which should be rectified by government intervention to limit the smoke—Pigou had imbued the positive science of economics with normative evaluation.

Coase made three moves in response. First, he argued that economic activities are not unidirectional but "reciprocal": Their effects always go in two directions. The smoke from the factory chimney, for example, would only have harmful effects on health if people chose to live nearby: Thus, "both parties cause the damage."[44] Conversely, to limit smoke, as Pigou had proposed, would impose a cost on the factory owner in the form of reduced production.[45] Why, Coase asked, should the factory have to accept the costs of reducing smoke for the benefit of the neighborhood? Instead, why shouldn't nearby residents pay the factory to reduce the smoke, or simply move elsewhere? Economists could not answer these questions, Coase argued, without imposing moral judgments inappropriate to a technical field. They could speak only to whether the value of clean air, assessed in economic terms, was greater or less than the value of the product that had generated the smoke. Second, following from this point, Coase argued that in highlighting the disparity between public welfare and private profit, Pigou had identified the wrong problem altogether. Only the "total social product," computed by weighing the gains of preventing a given activity compared to those of allowing it to continue, was relevant.[46] The goal was not to eliminate smoke altogether: To allow *any* claim of harm to prevent a smoky factory from operating might make everyone worse off. Rather, the goal was to achieve the "optimum amount of smoke pollution," defined not in terms of social welfare but as the "amount that will maximize the value of production," as determined by negotiations among producers themselves.[47]

Finally, the mere fact that some externalities were uncompensated was not in itself a sufficient argument for state intervention. State action came with transaction costs of its own, which might be more significant than those of either doing nothing at all or leaving the interested parties to work it out for themselves.[48] In instances where state action was warranted, moreover, the blunt and inefficient tools of taxation and regulation were not the only options.[49]

Instead, Coase argued that "the right to do something which has a harmful effect (such as the creation of smoke, noise, smells, etc.) is also a factor of production": Thus the state should assign rights to these activities, as it did to other factors of production, and allow private individuals to work out the value of smokeless air for themselves.[50] Rights, in other words, could be allocated by markets, just like any other good. If a producer wanted to generate smoke, they could simply pay the person harmed for the privilege, or vice versa. Regardless of who initially owned the rights to pollute, Coase argued, they would be allocated in whatever way maximized the total value of production.[51]

Markets and Morals: Comparing Views

Coase's analysis was rapidly embraced as a response to the framework of "market failure": redescribed by Chicago School economist George Stigler as the "Coase Theorem," it would become the far-reaching basis for a new approach to externalities.[52] Coase's was hardly the final word: Externalities would become a central concept in the fields of environmental and ecological economics, which emerged entirely within the six decades since "The Problem of Social Cost," as well as in the vast literature on public and common pool goods.[53] Yet the basic frameworks outlined by Pigou and Coase remain the dominant ones for thinking about the externality today. Pigou's analysis informs policies like the carbon tax and estimates of the social cost of carbon; Coase's underpins the likes of carbon markets and cap-and-trade policies. Despite important disagreements, discussed in depth below, they share most basic premises: Both reflect broadly liberal views of the roles of state and market from the vantage point of neoclassical economics. They assume that markets should generally operate without state intervention, and that the goal of intervention, where it does occur, is to restore market function, against proposals that the state set targets for industry to meet or prohibit certain substances or processes outright.[54] They assume, too, that individuals are the basic unit of market action, and in turn, of economic analysis. Both, in other words, accept many of the core principles of mainstream economic thought—neither was what we might now think of as an ecological or environmental economist—which makes it all the more striking that both wrote extensively about what we now tend to think of as environmental issues. Coase's examples in particular are teeming with nature: a field overrun with rabbits, cattle that stray from a rancher's field into a farmer's, a train whose sparking engine causes nearby woods to catch fire, a polluted stream with sickly fish, a building blocking the wind that powers a windmill. Externalities are not limited to "environmental" cases—yet they seem to reveal something about the effects of human action in a material world.

Where Pigou and Coase differ most significantly is in their accounts of what markets ought to do and where they fit in the broader social order—a difference that reflects a broader transformation in the political dimensions of economic thought. Pigou's attempt to merge utilitarian and marginalist thought echoed the eighteenth-century synthesis of commercial republican and market freedoms in its expectation that markets can and should realize the common good (now described in the guise of "social welfare") and its conviction that unintended harms to public wellbeing could be collectively tallied and addressed by a public entity. Coase's account, in turn, reflected the vision of markets that would become most prominent in the late twentieth century, which banishes the common good—by then suspiciously totalitarian to many liberals—except insofar as it shakes out in competitive markets ("total value production" as a Pareto-efficient outcome). His insistence that social costs must be negotiated reflects a view of the market as a force for pluralism, allowing people to make their own choices about values.[55]

It is perhaps not surprising that moral philosophers and political theorists have tended to find Pigou's account more appealing. Though Pigou himself is rarely referenced outright (other than in perfunctory nods to "Pigovian" taxes), his influence is visible in frequent references to the "true costs" of pollution or "social costs of carbon"; even in the idea that pollution obviously does constitute a "public harm."[56] Coasean markets in pollution, by contrast, typically come in for the sharpest critiques of economism's creep. Michael Sandel, for example, argues that paying for the right to pollute is troubling insofar as it suggests that there is nothing morally wrong with pollution—that it is "simply the cost of doing business, like wages, benefits, and rent."[57] Rather than simply condemning the Coasean approach, however, I am interested in what it reveals. For Coase *is* a more perceptive analyst of the externality than Pigou. He is right that Pigou's analysis relies on an unspoken and unjustified moral framework: To know that the market has failed to achieve optimal welfare, one must know what the optimal welfare is; to correct prices, a benevolent administrator (or moral philosopher) must know what they ought to be. Coase is right, too, that "social costs" are reciprocal and antagonistic—that one person's harm is another's benefit. And he is right to argue that harms like pollution are, effectively, factors of production, insofar as the transformation of some materials into new forms inevitably produces forms of excess matter. He is right, in other words, that Pigou and his followers take the meaning of social cost for granted and arbitrarily apply a normative standard to pollution—one that they typically do not apply to other kinds of economic goods. But it is not clear that such a stark distinction can be drawn: pollution, after all, is not *produced* by the emergence of markets in pollution rights, but by markets in standard commodities like cars and televisions—goods that most moral critics seem to think are legitimately bought and sold. If we recognize that "environmental" goods and bads are continuous with other kinds, then, we have two options. One is to claim that market mechanisms are appropriately applied to environmental problems.[58] The other is to focus critique not on *exceptions* to the rule of the market, but on the rule itself.

This is what the rest of this chapter does. First, it takes up a question that is largely missing from mainstream theories of externalities: It argues that by focusing on the rules of exchange alone, economists neglect the social relations of power in which capitalist markets are constitutively rather than contingently embedded. It then turns to address the dynamics of agency within markets, showing that markets *constitutively* detach intentions from consequences, at both the individual and collective levels; and render us unable to exercise the genuine responsibility that freedom entails.

Two Dimensions of Domination

Structural Domination: Social Costs in Class Society

Most analyses of the externality treat "markets" in the abstract, imagining individuals who meet as formal equals and enter into voluntary agreements based on a rational assessment of their options. In markets in capitalist societies, however, people come to exchanges from structurally unequal positions of power. This is the central insight of a less well-known twentieth-century theorist of the externality: the German economist K. W. Kapp, an institutionalist informed by the Frankfurt School of critical theory. In *The Social Costs of Private Enterprise* (1950), Kapp described not the widespread affluence typically thought to characterize the postwar period, but an economy characterized by workplace injury, polluted air and water, depleted plant and animal resources, and mounting waste—all, he charged, costs of production paid not by private industry but by "society."[59] Entire industries were only profitable, Kapp claimed, because they had managed "to shift a substantial part of these costs to other persons and the community at large."[60] Cost-shifting, he thought, was more pervasive even than exploitation, such that capitalism itself was "an economy of unpaid costs."[61] For Kapp, this meant that social costs were a site of political conflict: Echoing Karl Polanyi's theory of the "double movement," he argued that political history since the nineteenth century could be read as a "revolt of large masses of people . . . against the shifting of the social costs of production to third persons or to society."[62] Private enterprise pushed costs onto society, and society pushed back.

What is striking is that the moral economist Kapp agrees with Chicago School lodestar Coase on two key points: that pollution constitutes an unpaid factor of production, and that social costs are reciprocal, insofar as the cost for one person is often a benefit for another. Yet for Kapp, these points are the basis of his critique that industry profitability has come at society's expense, and that social costs are a site of struggle. This unexpected agreement presents a puzzle. If Kapp is right that many industries are profitable *because* they shift costs onto others, then treating the right to pollute as a factor of production that must be paid for, as Coase proposes, would seem to give society a tool in the struggle to shift costs back. If forced to pay their full costs, Kapp's argument suggests, many industries might be forced to drastically reorganize production or even cease it altogether; capitalism itself might collapse.[63] If companies had to pay the IPCC's proposed carbon taxes, for example, ranging from $135 to an astonishing $5500 per ton, many would likely no longer be viable.[64] Coase, by contrast, argues that while the right to pollute should be treated as a property right, it doesn't matter who initially holds it.[65] The relevant parties, he claims, will simply negotiate an agreement that maximizes the total value of production, inclusive of any necessary compensation for harm. One of them, it would seem, has to be wrong.

While Kapp's argument foregrounds political conflict, it is vulnerable to the same critique that Coase had launched at Pigou, insofar as it operates on the premise that there *is* a "true cost" to pay, one that could capture the various *in natura* harms to human and nonhuman life. For Coase, by contrast, treating pollution as a factor of production means that there are no true costs—there are only the prices that the relevant parties negotiate. (The point of Pigovian taxes, moreover, is not that people will actually pay the full cost in perpetuity, but that the increased cost will drive a shift to or creation of alternatives.[66]) On these grounds, Milton Friedman would later argue that societies should permit "only pollution that's worth what it costs, and not any pollution that isn't worth what it costs"—and that the only way to determine which pollution was "worth it" was to let individuals themselves decide. This, both Coase and Friedman thought, could happen even absent formal rights to pollute: People might choose to live near a smoky factory if rents were lower; countries entering into trade agreements could decide what level of pollution to allow. Japan, for example, could choose to produce steel and accept the ecological costs, while environmentalists in the United States could be happy with cleaner air and water: "If Japan chooses to subsidize the export of clean air to the United States," Friedman asked, "why should we object?"[67] This particular example isn't entirely applicable to climate change, of course, insofar as carbon emissions anywhere are a problem for people everywhere. But I want to focus on Friedman's question: Why *should* we object if some people choose to accept the costs associated with production in order to realize the benefits? After all, this is precisely the trade that many developing countries have made in the years since.

Let me pose the question more provocatively. What's wrong with letting the lowest wage countries choose to accept a trade in toxic waste? If "underpopulated" African countries want to import "visibility impairing particulates" and export "pretty air" to the United States, why should we object? This, of course, is a rephrasing of

Larry Summers's notorious defense, in a 1991 World Bank memo, of the "impeccable" economic logic of "dumping a load of toxic waste" in low-wage countries, especially those in Africa.[68] Summers was widely castigated for his remarks. Yet the economic logic *is* impeccable on its own terms: this is precisely the logic of comparative advantage, applied to environmental regulation.[69] If we object to the statement—as I think we should—we need an answer to Friedman's question. Debra Satz offers one: For Satz, markets in toxic waste are a "noxious market" characterized by severe power imbalances or weak agency, or that result in "extremely harmful outcomes," either for individuals or society.[70] Satz is right that we should be troubled by the noxiousness of markets in toxic waste. But we should be especially troubled by how many markets turn out to be noxious. The severe inequality that concerns Satz is in fact the norm rather than the exception: It is the constitutive basis of class society itself.

This is what Kapp saw most clearly: that the struggle over the burden of social costs is better characterized in terms of struggle between classes with disparate power than as a market exchange between equal individuals. Producers responsible for social costs, he argued, had the upper hand. This was in part due to the nature of social costs, which were typically diffuse while benefits were concentrated. It was especially true in instances where the visibility of costs lagged behind the realization of benefits, sometimes taking years to emerge, as in harms to human health; or in instances in which the costs of private production simply appeared to be forces of nature. Economists tend to lump these kinds of challenges to collective action under the banner of transaction costs; the Coase theorem's assumption that these costs are nonexistent is one of its well-canvassed shortcomings.[71] But the problem Kapp identified was not only the general difficulty of organizing individuals to act together or the particularities of environmental effects: As he noted, private producers typically had financial and organizational resources outstripping those available to those who bore the brunt of social costs, and, in the absence of organized opposition, could easily impose them on others. To put it differently, although the factory owner and members of the neighborhood may be formally equal before the law, they stand in radically different positions of power.

Consider here G. A. Cohen's example of a town in which a chemical company opens a factory, offering jobs that pose serious health risks.[72] Do the factory's workers freely choose these jobs, he asks, or are they forced to take them? Those who argue the latter emphasize the structural conditions of labor under capitalism, in which those who lack property are forced to work for the owners of capital. Those arguing the former tend to make a version of Friedman's argument: If someone is willing to risk their health for a better wage, why should we stop them from doing so? If someone doesn't want to take a given job, they can look for another: They are not bound to any particular employer. The fact that they face dismal options or material hardships does not diminish their freedom to choose between them.[73] Cohen concludes that the worker is at once freely choosing the dangerous job *and* severely restricted in options. The worker needs a wage to survive in the near term, even at the potential cost to life in the long run; it may very well be that their best option is to take a health-threatening job, while others are able to "make money out of [his] relative lack of freedom."[74] This may be unjust, Cohen notes—but it is not necessarily illegitimate on capitalism's terms, insofar as it is the result of a formally free exchange.

A critique, then, requires a challenge to those terms. One comes in the observation that capitalism is foundationally structured by a division in social power. People who own nothing sell their laborpower to those who own the means of production; when the latter pay for labor, they purchase the power to direct the worker's bodily capacities—to direct the worker's literal *body*—for a given period of time. At the very root of capitalist production, then, is the expectation that some people will regularly submit to the will of others. The legitimacy of this authority rests only in the fact that workers have, at least ostensibly, freely chosen to enter into such an agreement. This presumption is what many critics have challenged.[75] Arbitrary power is at work, Alex Gourevitch argues, in the relationship between classes in which some people own while others work. Differences in resources, in these circumstances, grant not only a wider range of consumer choices, but the ability to compel others' activity.[76] The unequal division of assets in capitalist societies, in other words, is not only a measure of injustice, but a mechanism of what Gourevitch calls "structural domination."[77] As

Nicholas Vrousalis similarly argues, "capital just is monetary title to control over the labour capacity of others."[78] This imbalance in power also means that laborers are structurally disadvantaged in negotiations: The capitalist will lose money if the factory lays idle, but if the worker doesn't work, they will struggle to secure their most basic needs. This means, put bluntly, that laborers often have to sell their labor for whatever price they can get. Similarly, someone who needs a wage to survive is likely to accept not only physically taxing jobs but physically unsafe ones, in which they are exposed to dangerous levels of smoke or toxins: They are likely, in other words, to accept a higher burden of social costs as a result of their structural position. In turn, their reliance on the market means that they will have to accept whatever kinds of social goods they can afford.

Most accounts of structural domination focus on the immediate relationship between labor and capital. But structural economic domination extends beyond the workplace. Capital is not only power over others' labor per Vrousalis, but power over investment and hence purposive action more broadly: over what is produced and how. It is a title to direct not only other people's labor but also other factors of production, including the right to pollute. It is the power not only over the production of commodities, but over the production of environmental conditions, and increasingly, over the condition of the planet itself.

Let's now consider a twist on Cohen's question: Are people who live next to factories with smoky chimneys *forced* to live there? They are legally free to live anywhere they can afford, after all; feudal title aside, no one is bound to a particular landlord. Most people, we can assume, would prefer to live where they can breathe clean air. But pollution is matter that no one wants to buy; that, instead, people pay to *avoid*. Those who cannot afford to avoid social costs are therefore likely to end up paying social costs *in natura*—in the form of asthma, say, or heart disease. Those who cannot afford to avoid the social costs of production are almost always those who are most dominated in the labor market; those whose only options are the lowest paying jobs and thus the lowest cost apartments. The wealthy live in leafy, upwind neighborhoods; the poor, next to incinerators and power plants. Although these particular examples pertain to localized environmental harms rather than the global phenomena of climate change, they are more closely connected than is sometimes suggested. The future harm caused by the accumulation of carbon molecules is often paired with the immediate and localized impacts of fossil fuel production and use, from car exhaust to refinery emissions to black lung disease; and in any case, despite some notable exceptions—Malibu mansions, Miami real estate—climate vulnerability tends to follow existing patterns of social vulnerability.[79] We could just as easily pose a question about whether those who live in flood zones are forced to live there. To lack money, in these instances, is not only to lack the ability to realize one's aims in the material world, but to lack the ability to *refuse* the costs imposed by others.[80] Distributive inequality (itself stemming from social inequalities of power) thus conditions not only welfare but freedom.

In fact, it is far easier to impose social costs on other people in the course of production than it is to compel their labor. To use someone's body for labor typically requires consensual exchange in some form (however limited) because agency rests with the laborer: If you ask someone to work without paying, they can simply refuse. When using someone's body as a sink for smoke or particulate matter, by contrast, the agency lies with the polluter. Someone can dump toxins in the water or carbon in the air whether or not they have obtained consent to do so. Those living nearby may protest—but they cannot refuse to breathe polluted air in the way that they can refuse to work; they cannot refuse staggered effects like a heat wave or flash flood.[81] Coase suggests that the right to pollute, if recognized as an aspect of property, might be allocated to either the producer of pollution or its bearer. But in actual fact the right to pollute lies with the polluter by default, and by default it is available for free. To override this default, moreover, is extremely demanding—and demanding of state action in particular.[82] This feature is rooted in the material qualities of pollution and would present a challenge for any form of social organization. But in a social order in which one group of people enjoys institutionalized structural advantage over another, it becomes a source of domination: a way that one group of people is able to profit by arbitrarily interfering with others.[83]

This structural domination would remain even if the right to pollute *were* formally treated as a factor of production, as Coase proposed. Say a state created a right to pollute and allocated it to residents of the neighborhood rather than the factory owner, such that the factory owner would have to negotiate with everyone living in the vicinity if he wanted to produce smoke. (Put differently, we might imagine granting everyone an alienable right to a clean environment.)[84] If each person was paid the "true cost" of emissions on their future wellbeing, calculated in terms of lost wages, health-care costs, shortened lifespan, and so on, the cost might well be staggering. (Although even in nonmarket forms of accounting and morally motivated evaluation, wages are often the metric according to which people's relative worth is assessed, such that it is always impeccably logical to impose costs on the poor.)[85] But in reality, people are likely to sell the right to pollute for far less than its "true cost." They might do so in ignorance, not realizing the likely long-term effects; or they might know that there will be long-run consequences but need money in the near term; or they might worry that the factory will move elsewhere—down the road, over the border, around the world. The reasons people might sell the right to pollute—in this instance, the right for someone else to impose the physical harms of production on one's body—more cheaply than they "should," in other words, are precisely the same as those that compel laborers to sell their labor cheaply: They are dependent on the market for their means of life, and they own nothing else. When people who have nothing to sell but their labor find that even that is not particularly valuable, they may decide that their competitive advantage lies in their willingness to accept particularly dangerous or dirty forms of production—precisely as Friedman and Summers suggest.[86]

As Summers's example well illustrates, moreover, structural economic domination is always articulated through other social relations: most notably, in these cases, those of race and nation.[87] In many instances, direct political domination and economic domination work in concert. When people are subjected to domination by the state, including forms of racial domination, they are often exposed to indirect domination by capital.[88] Waste facilities, for example, are most likely to be sited in communities that are unable to muster effective political resistance, which in turn are often those that are least represented within the state or even treated hostilely by it. In the United States, this means that waste facilities are located disproportionately in communities of color, as the environmental justice literature has exhaustively documented.[89] In other cases, people's lack of formal political standing exposes them to extreme forms of economic domination, as when the dirtiest and most dangerous forms of work are performed by undocumented migrants who lack claims to state protection.[90] In still others, the lingering effects of direct domination are reflected in contemporary structures of economic domination: Formerly colonized countries bear a higher burden of pollution and are more likely to accept trades in toxic waste; they are more vulnerable to the effects of climate change.[91]

This is not an exhaustive account. The central point is that the distributive dimensions of climate change are not only matters of injustice but of unfreedom. It is undeniable that there are stark and disturbing disparities in the forms of consumption that generate harmful forms of matter, of which carbon emissions—correlated almost perfectly to income—are only the most obvious example.[92] The goal of a more egalitarian distribution of both "goods" and "bads" is an eminently worthy one—even if it is not, I think, all we can hope for. Nevertheless, we should be concerned not with the unequal distribution of bads and goods, but with the constraint on *genuine* freedom of choice rooted in an inequality not only of resources but of power—an inequality that is not contingent but necessary to capitalism as a social order.

Freedom in the Market? Market Domination and Nonresponsibility

If the account offered above addresses disparities in the costs and benefits of emissions, it may seem unsatisfying in other respects—perhaps particularly so with respect to climate change. It is clear that certain industries, most notably but not exclusively those dealing in fossil fuels, bear outsized responsibility for climate change, especially insofar as they have intentionally delayed political action.[93] But focusing exclusively on private industry and investment seems to ignore the fact that carbon emissions, as well as many other environmentally relevant externalities, are produced by many individual actions. If we are honest, many would argue, we must admit that everyone who consumes carbon-intensive goods—which is to say, nearly everyone presently living in the industrialized West—is complicit. This is the fundamental premise of a major philosophical body of work focused on responsibility for climate change, and on the particular challenges that this phenomenon poses to familiar accounts of moral agency. Climate change is caused by the actions of such a large and diffuse number of people that any individual's contribution seems impossible to parse: Agency is, in Stephen Gardiner's terms, "fragmented."[94] At the same time, each individual action is so small that it seems to escape responsibility. Indeed, climate change seems to reveal what Judith Lichtenberg describes as "new harms."[95] Once, Lichtenberg suggests, we could recognize the actions that caused injury and seek to avoid them, even if the moral questions they raised were complex; but when our most mundane activities turn out to contribute to severe harms, the fundamental calculus shifts. As Lichtenberg observes, "Not harming people turns out to be difficult and to require our undivided attention."[96] It requires reevaluation of our seemingly trivial choices: what we eat; where we live; how we get around; what we wear.

In light of this realization many philosophers have sought to link causal responsibility more clearly to moral responsibility, drawing lines between the activities that physically emit carbon, however minor the quantities, and their morally problematic effects on others, however distant. While any one person's contribution to climate change may be negligible to the point of imperceptible, many philosophers suggest, to ignore the aggregated effects of individual actions is to make what Parfit called "mistakes in moral mathematics"; we each have a duty to minimize actions that, when combined, add up to a serious harm. Getting serious about climate change means facing up to our own contributions to the problem.[97] Tellingly, in these discussions, action often amounts to consumption: iPhones, flights, steaks, SUVs. As Lichtenberg puts it bluntly, "Every bite we eat! Every purchase we make!"[98] This suggests that rather than constituting a source of "new harms," climate change has simply shed new light on an existing class of harms: those already embedded in our market choices, if only by exclusion from them. Yet although nearly all of the actions in question are mediated by the market, this crucial social institution goes almost entirely undiscussed. This is a remarkable omission. Why is it that moral agency has come to be located so significantly in consumer choices? And how might we understand responsibility differently if we attended more closely to the way markets structure the relationship between intent, action, and outcome?

Worries about climate complicity are only the most recent in a long train of anxieties about responsibility for unintentional and often unwitting harms done to others through consumption, from "slave sugar" to "blood oil." In this, critiques of complicity often carry a localized echo of Marx's critique of commodity fetishism, which enjoins us to recognize the relations of domination lurking within the mere things we buy every day. The problem of complicity is often framed in epistemic terms: Because the connections between our seemingly innocent commodities and the horrifying circumstances of their production are difficult to see, they must be made visible. Often, this means directly equating the act of making a purchase to harm or violence: "If we purchase the commodity," the abolitionist William Fox insisted, "we participate in the crime."[99] But suggestions that the relationship between consumption and harm is direct can obscure rather than illuminate the material and structural forces that shape and channel our actions, insofar as they achieve this moral immediacy precisely by eliminating the mediating social relation in question. It's important, in other words, that these actions are *not* immediate, but mediated by the market: The harms that result from our purchases are *not* the same as those we enact on others directly.

Theorists of moral agency and climate change often emphasize the difficulty of knowing how to act rightly: Lichtenberg, for example, suggests that to avoid being complicit in harms requires us to do an immense amount of research about each and every one of the things we buy. But there often *is* something genuinely tricky about locating harms in market choices: namely, that it really *is* hard to see, on the surface, what is wrong with them. Their bad effects seem to emerge from interactions that are, on their face, unobjectionable: the kinds of ordinary, legal market transactions that we make unthinkingly every day. While some goods, like slave-made sugar, are obviously morally objectionable, in other cases there is nothing especially wrong with the particular good in question: It is only in the extreme aggregate that problems like climate change emerge, for example. Avoiding complicity in "new harms" would thus seem to require an immense amount of research about each and every one of the things we buy. But this expectation fundamentally mistakes the structure of action in the market. Epistemic problems are not exceptional but endemic to markets. As Leif Wenar puts it, "we simply cannot know exactly which products are tainted by moral toxicity in their supply chains"—even as we know that "the taint is there."[100] Market coordination is premised on our ignorance. By their very nature, markets systematically make the negative externalities of our consumption decisions opaque to us.

Prices are a thin kind of knowledge: They simply aggregate information about supply and demand, so that you don't need to know anything about where the lead in your pencil came from or the soil where your bananas grew. This means that for the most part, we have no way of knowing about externalities: about what kinds of noxious matter are generated as byproducts of the commodities we buy. We may be aware of certain high-profile examples—the carbon emissions associated with flying, for example, or the deforestation that results from expanding ranches for beef cattle—but there are countless others we don't take into account because we do not, cannot, and are not expected to know that they occur. It is not a problem, according to the theory of market freedom, that we don't know these things—to the contrary, it is a boon. For Hayek especially, it is precisely the limits of our knowledge that recommend the market: The "marvel" is that it coordinates individual actions on the basis of the minimal information contained in the price.[101] In the conventional wisdom, externalities are presented as simply an epistemic problem: Missing prices means missing information. Once externalities are internalized, morally troubled consumers can return to their previous state of blissful ignorance. One problem is that the prospect of internalizing *all* externalities, or even having all the information necessary to do so, is fantastical: It relies on the idea that all costs can be identified and incorporated into prices; that the market can provide a 1:1 model of the world in its entirety, down to the last carbon molecule. But even bracketing this issue, the difficulty for those concerned with moral responsibility is that market choices, particularly those made under conditions of market dependence and competition, remain within a framework that fundamentally denies it.

The market, for thinkers like Hayek, is a way of acting cooperatively but not socially: People make free choices as individuals, and those individual choices are aggregated through what Smith called the "invisible hand" and Hayek calls "spontaneous order." Crucially, markets aggregate individual actions *regardless* of intent. In other words, they detach intentions from consequences. Eric MacGilvray therefore argues that a fundamental principle of market freedom is *nonresponsibility*: It consists in the "ability to decide for oneself how to respond to the menu of choices that one faces without being publicly accountable for the consequences of those decisions, and thus with the ability to impose certain costs on other people without their consent."[102] An avocado fad may send farmers' fortunes rising and plummeting; our propensity to save rather than spend may plunge workers into unemployment—but no one can be said to be at fault. No one is directly responsible for the prices of goods; no one is directly responsible for the distribution of income. In this, externalities are a *reflection* of the general rule of the market rather than the exception to it: As MacGilvray argues, *"market prices themselves are externalities,"* insofar as they "impose costs and confer benefits on third parties in ways that no one—least of all the affected people themselves—can predict or control."[103]

Condemnatory though it may sound, the charge of "nonresponsibility" is not, for MacGilvray, a critique. To the contrary, this disconnect between intention and result has often been a point in markets' favor. It is key to the optimistic view of unintended consequences reflected in ideas of the "invisible hand" and "spontaneous order."[104] Individual actors need not be enlightened or altruistic, wise or kind, these accounts proclaim; they need not consider the effects of their actions on others or anticipate the future. To the contrary, it is expected that they are not, and will not. The happy fact, Smith thought, is that these self-interested actions will nevertheless turn out for the best.

This uncoupling of intention and result is also why commercial society theorists and neo-republicans have often found the market appealing as a counter to the arbitrary power of individuals.[105] Even where markets bring about detrimental outcomes, it is no one's intent and no one's fault that they do so. No single person is responsible for a rise in grain prices; thus no single person dominates someone who cannot afford to buy bread. If no one is responsible for the effects of their actions on anyone else, no one can exert their arbitrary will over anyone else. In this sense, Philip Pettit suggests, the market is akin to a force of nature—and as such, it is not a force for domination.[106] Markets have therefore often seemed to pose an answer to core problems of political thought: a form of order that takes men as they are and not as they might be; that realizes the common good treasured by the ancients while protecting the individual liberty prized by the moderns. For some thinkers, even the idea of common good has dropped away. Hayek's spontaneous order, like the "invisible hand," also reflects a view of unintended consequences as "felicitous," as Daniel Luban notes—but it strips away the normative inflection of something like Mandeville's "public virtue."[107]

When Hayek proclaims that liberty and responsibility go together, then, this is not responsibility as many moral philosophers imagine it—responsibility for the effects of our decisions on other people, let alone for the direction or shape of collective life. Markets *are* supposed to allow us to make choices about what we as individuals value; and more than this, to force us to make choices about our real commitments when faced with material constraints.[108] Yet if we are responsible for our own choices and their effects in our lives, we are not therefore responsible for the effects of those actions on others. If someone chooses to live near a smoky factory so they can pay less in rent, why should we prevent them from doing so? This view of responsibility, in other words, is perfectly compatible with what MacGilvray calls nonresponsibility: it does not, according to Hayek, make us "accountable for our actions to any particular persons."[109] This, of course, diverges sharply from the view that we bear moral responsibility for the effects of our consumer purchases: Market freedom expresses precisely the opposite idea. So, too, does Hayek's view that the limits of individual knowledge recommend the market as a way of aggregating the knowledge of all cut against the idea that we could simply research our way to moral choices. If we could know everything that went into our purchases and calculate the best outcome for everyone on earth, we wouldn't need markets at all.

If markets cut against individual responsibility, they cut still harder against collective responsibility. As a rule, capitalist markets limit the scope of our choices and ability to act on our judgments—including our judgments about what would constitute appropriate, respectful, or reciprocal relationships to one another and to nonhuman nature. People do retain a degree of agency about what they choose to buy and sell. But those choices are deeply limited in their own right, and coordinated through a market that not only aggregates but compels countless individual decisions in ways that defy both individual and collective control. However much we agonize over our consumer choices, the market is not and cannot be a deliberative space or a site for the exercise of collective reason. This is concerning—for while we do not all need to agree on a common good, we do need to be able to work toward the common purpose of maintaining a habitable planet. Organizing society so centrally around a mechanism whose very purpose is to divorce action from effect, however, threatens to abandon the possibility of common purpose and self-rule altogether.

If the republican critique of domination as arbitrary power has been potent in diagnosing power exerted in the workplace, and can be extended to address relationships beyond the workplace as discussed above, it is less well-suited to these diffuse forms of compulsion and impotence. The market does away with the qualities of interpersonal and intentional action central to the republican concept of domination.[110] These are actions that no one in particular forces one to take; outcomes that no one intends; processes that seem to unfold "behind our backs." The most robust defense of a republican account of domination by the market comes from William Clare Roberts, who argues that because the market does subject us to other people's choices and actions, even if indirectly and unintentionally, the domination specific to capitalism is *inter*personal even if *im*personal. The market is, as Roberts puts it, "just people."[111] But this stretches the republican concept of domination to the breaking point—while failing to adequately challenge the account of the market given by its defenders. Hayek himself openly grants that the market is often out of our control, but nonetheless holds that the "impersonal and seemingly irrational forces of the market" are preferable to the "equally uncontrollable and therefore arbitrary power of other men."[112] However more appealing one might find Roberts's account of impersonal domination, Hayek's is more obviously consistent with the neo-republican view of freedom.[113]

Markets as Seriality

The quality of this unfreedom—what I describe as market domination—is better captured by Jean-Paul Sartre's concept of seriality.[114] Before discussing this particular concept in more detail, I want to explain the turn to Sartre on this point. In my view it is precisely the resonance between Hayek's and Sartre's views of the freedom—and duty—to choose that makes Sartre's analysis of the constraints on that freedom a powerful rejoinder to Hayek. For while Hayek's insistence that one person must not be subjected to the arbitrary will of another is easily read in terms of republican liberty, there is also something more like self-determination in Hayek's view of what markets do.[115] Markets, for Hayek, give us the freedom to choose what we as individuals value—and more than this, they force us to make choices about our real commitments in the face of material constraints. We can assert any number of principles in the abstract, but the things to which we choose to devote our limited resources reflect our *real* commitments, far more than any mere statement of belief.[116] Precisely by virtue of its limitations, he argues, the sphere of market choice is "the air in which alone moral sense grows and in which moral values are daily re-created in the free decision of the individual."[117] Hayek insists that the "opportunity to build one's own life also means an unceasing task, a discipline that man must impose upon himself"—and this, he claims, is why "many people are afraid of liberty."[118]

The resonance with Sartre's early existentialism is striking. "My freedom is the unique foundation of values," Sartre declares; "*nothing*—absolutely nothing—can justify my adopting this system of values rather than that one."[119] We alone are responsible for our choices, Sartre emphasizes, and this responsibility is hard to face. It leaves us with no excuses. This, of course, is the basis for Sartre's charge that freedom is anguish; that we are condemned to be free. While to some this echo may seem suspicious, I draw on Sartre to recover a view of freedom as choice and responsibility that is *not* automatically located in the market. What is most powerful in Hayek's account of freedom, in my view, is precisely its intimation of an anti-foundationalist approach to value, and its insistence that we have no choice but to choose. It is only by recognizing that values *are* contingent rather than given, after all, that we can freely choose them; and only because they are *our values* that we care if they are overridden. The paradox is that if the market limits the arbitrary power of individuals and collectives to assert their will over others, it also limits the power of each person as an individual to make decisions that reflect their own ends, as Hayek insists we ought to be able to do. Existentialist accounts of the freedom to choose our values, as I discuss in more detail elsewhere, can help reveal the *inadequacy* of freedom in the market, without undermining the value of choice itself. Market freedom, on this view, is a perversion rather than realization of existentialist freedom, insofar as it undercuts our ability to be genuinely responsible for our decisions.[120]

Here, let's look more closely at Sartre's analysis of seriality. Outlined in *Critique of Dialectical Reason* (1960), a series, for Sartre, is a loosely constituted collective of individuals aggregated around structures or practices to which all bear the same relation. Members of a series are often aware of one another, but their relationships to one another are impersonal and detached. Their actions adhere to the same practices and rules, but they are undertaken in relative isolation. Because they are interchangeable, members of a series typically view one another as potential competitors. Markets, for Sartre, are a prime example of "seriality as a force which is suffered in impotence." The market "has an undeniable reality" that is "imposed on everyone," but that no one freely chooses.[121] This imposition is of a distinctive kind. Although sellers are in a competitive relationship to one another, theirs is not the antagonism of direct struggle, in which the winner gains the power to impose conditions directly on the loser. It is not exemplified by the classic republican case of a master who exercises arbitrary power over a slave; it is, rather, a situation in which *everyone* is the abstract other *to everyone else*. People act in anticipation of others' actions, often hastening the effect they anticipate in the process: for example, when people dump stock that they fear will decline in price, they help to bring about its crash. This anticipatory relationship is in some ways reminiscent to that described by republicans, in which someone who is subject to the arbitrary power of another acts in servile expectation of the more powerful person's desires and commands—but the difference is that in the market, there *is* no actual person giving commands. The relationship between any given buyer and seller is conditioned by relationships among thousands, millions, billions of other buyers and sellers, all expressed through the medium of price. We therefore act in anticipation of the actions of anonymous others who we do not know and will likely never meet; people who have no direct power over us, and never will—and who themselves are acting in anticipation of us.

Within standard economic theory, decisions in the market are said to reflect our preferences—at least, the preferences of those who have the money to express them. Those who don't are largely invisible to the market; in this, the poor, future generations, and nonhuman species are in more or less the same boat. But even those who *do* have money cannot simply choose what value they attribute to different goods. Although I may well think clean air is worth quite a lot, in a true market I cannot pay what I think something is worth, or even what I negotiate with another individual. I can only pay what the market—as the aggregation of millions of other buyers and sellers making transactions in competition with one another, under various constraints and forms of domination—decides it is worth. In the language of economics, we are all price takers, rather than price makers.

Even when we have firm moral judgments about certain forms of consumption, moreover, we will often find it difficult to act on them in light of the pressure of price itself, making cheaper purchases even when we know that they are morally worse.[122] If a train ticket costs twice as much as a plane ticket, I will be tempted to buy the latter, even though I know it is more environmentally harmful.[123] While this pressure bears on everyone, it is particularly acute for those whose wages are closest to the costs of reproduction. Those who rely on wages must make those wages stretch as far as they will go—and because workers compete against one another for jobs, those who can cut the costs of their own reproduction will often find themselves at an advantage in the labor market, able to offer their labor power for a lower cost. If I am so committed to buying fair-trade goods that my cost of living is much higher than that of my peers, I will find myself at a disadvantage in the labor market. Some people, of course—likely many of those reading this chapter—have enough disposable income to direct toward satisfying moral preferences: paying extra for cage-free eggs, say, or sustainably produced clothing. But these are not the conditions under which most people live today.

The point is not simply that all consumption is equally problematic or that the wrong life cannot be lived rightly.[124] We do retain a degree of moral agency even within severe constraints, and should take care with our choices—especially those of us who can do so without serious hardship. But the personal choices of a small group of elite consumers are hardly the core questions of politics. What is far more significant is the manner in which our actions are channeled and constrained in ways that are at once largely beyond our control and systematic in their effects. We should be acutely aware not only of the limits to our efforts to enact change as individuals, but of the ways that our individual choices *themselves* are limited; indeed, the very fact that nearly all of our action takes place in and through market purchases is itself a sign of such limits.

We should be aware, too, of the ways that market domination pertains to many forms of collective as well as individual action. Governments at levels ranging from the municipal to the national compete to attract private investment that can supply jobs and income for residents, and sources of tax revenue to fund their own operations. To do so they may relax environmental regulations, fast-track permitting processes, or even actively entice polluting industries.[125] (This is all the more likely in industries whose most severe effects are often felt elsewhere, as is the case for many carbon-intensive industries.) The more dire the straits of the political community in question—perhaps they are suffering the aftermath of deindustrialization or deeply indebted following an externally imposed structural readjustment—the worse the options will be. In such instances, the owners of capital are positioned to make money from others' dependence on their accumulated resources.

Even capitalists themselves, moreover, do not act freely according to their intentions. Hayek argued for markets against states on the basis that individual freedom "cannot be reconciled with the supremacy of one single purpose to which the whole society must be entirely and permanently subordinated."[126] Markets do not themselves subordinate people to a single purpose—but capitalism does. Capital's sole purpose is self-valorization: Private investment happens only to the extent that it is expected to be profitable, as determined in the field of a competitive market. A capitalist who fails to receive a return on their investments will not last long. Thus, capitalists tend to dump waste as cheaply as possible, just as they seek to obtain labor as cheaply as possible—not because they are cruel or greedy, but because they themselves are dominated by the imperative to produce value and disciplined by market competition. Of course, companies sometimes gesture to competitive pressures as an excuse for why they simply *cannot* improve conditions, and as Madison Condon's comments in her chapter in this volume remind us, we should be wary of instances when these excuses are offered in bad faith. It is often strategically and politically effective to target major corporations and sectors that have outsized market power relative to individual consumers, as in campaigns to divest from fossil fuels. Yet structural pressures are real, and analysis demands that we confront them. Few corporations will pay to install costly and unproductive pollution reduction technologies or reduce their carbon emissions if their competitors do not—as corporations themselves have repeatedly emphasized.[127] If identifying structural incentives threatens to let powerful companies off the hook, focusing on corporate malfeasance alone threatens to suggest that systemic problems can be solved by reforming a few bad actors.

Across different levels of action, then, markets constitutionally disaggregate intent and consequence, frequently compel us to act against our own better judgment, and limit our ability to choose values both individually and collectively. Perhaps this disconnect between intentions and consequences would be less consequential if we had faith in Smith's and Mandeville's auspicious view of their ability to produce happy ends. But as climate change makes all too clear, unintended consequences are just as likely to be perverse as fecund.[128]

Conclusion

As suggested at the outset, my argument in this chapter is not that climate policies that rest on Pigovian or Coasean frameworks are, ipso facto, unacceptable or unworkable—although I am skeptical of the more optimistic claims made on their behalf. Carbon taxes and cap-and-trade programs may be useful tools in a broader program for decarbonization—although the gap between the average price of carbon emissions (roughly $8/ton) and the price that the IPCC suggests would keep temperatures below 1.5°C (estimated between $134–$5500/ton) is sobering.[129] As Kysar rightly notes in his chapter, moreover, the emphasis on undertaking an energy transition at least cost rather than maximum speed itself betrays troubling priorities.

It is nevertheless possible, in the realm of ideal theory, to imagine the use of carbon prices to smoothly wind down the use of fossil fuels. The elegance and seeming precision of this solution has long been key to its appeal among economists—and perhaps among moral philosophers as well. As Kapp's analysis rightly suggests, however—and Coase's also acknowledges—the shape of climate policy will ultimately reflect struggles over power more than philosophical insights. After all, Coase is right that there are no "true costs," "correct prices," or "objective values." Cost, price, and value must be seen as sites of politics—not individualized consumer politics or Hayekian expressions of individual freedom, but collective judgments about what we value and what ends we want to achieve, which are refracted through conflicts across radical power imbalances. In other words, it is necessary to bring the problem of social cost into the realm of political economy and collective action rather than the realm of individual moral action or the technical calculation of "true costs," whether by ecological economists or moral philosophers. "Social costs" are more like wages to be struggled over than rates to be calculated.

Indeed, as Condon observes, the political failure of carbon taxes and carbon pricing has, in recent years, prompted a turn away from a focus on market mechanisms alone toward more robust state intervention in service of decarbonization, as reflected in the 2022 Inflation Reduction Act.[130] But this does not mean that, as Condon suggests, climate change is not actually an instance of an externality. While programs like the IRA seek to lower the cost of renewable technologies rather than raising the cost of fossil fuels, using "carrots" rather than "sticks," the basic problem of harmful effects that are not reflected in prices remains as persistent as ever. Where I unequivocally agree with Condon is in her suggestion that many climate policies have neglected the entrenched power of polluters.[131] Although fossil fuel producers have often supported carbon taxes when more aggressive decarbonization measures loomed, after all, they have never supported anything close to the level of carbon pricing that would be required to eliminate fossil fuel use, since this would eliminate their primary product. Indeed, the shift toward models of climate policy that demand more robust state intervention—in some instances including measures aimed at addressing the social disparities that inflect both the causes and effects of climate change—itself reflects the political pressure brought to bear by social movements.

Debates over carbon pricing are ultimately but one illustration of a broader set of challenges that arise when capitalism organizes human action on a planetary scale. Some of the problems raised in this chapter exceed capitalism, of course. Human relationships will always be mediated by the material world; our actions will always have unintended consequences on other people, and on the nonhuman world.[132] The intense difficulty of assessing responsibility in a world where our most mundane actions inevitably ripple outward, with consequences that exceed our intent or control, will remain. So too will the spatial difficulties of acting within bounded communities when the effects may have planetary repercussions, and the temporal problem of actions and effects that span generations. It nevertheless remains vital to understand these persistent features of action with reference to the specific social relations that constrain and direct it.

Externalities, which point to the impossibility of limiting the effects of action to the parties to a contract, pose a genuinely challenging political problem. As Sharon Krause argues, "true noninterference is not an option for any of us in the environmental domain"; the elimination of even the *possibility* of arbitrary interference is even less imaginable.[133] Rather than abandoning the concept altogether, however, this is precisely why it is crucial to address domination at a *structural* rather than individual level: the ways our actions are channeled, organized, and even compelled; the ways our relationships to one another are ordered; and how these relationships are both structured by and themselves shape the material world. It is crucial, in other words, to consider what Young calls the "institutional conditions which inhibit or prevent people from participating in determining their actions or the conditions of their actions"—as well as the way that the physical effects *generated* by institutional conditions inhibit or prevent people from living freely.[134] As the concept of the externality discloses despite itself, human action inevitably has material consequences, many of which unfold in unexpected ways. But the dimensions of market and structural domination I have addressed above mean that capitalism constitutively produces them, that it produces them differentially, and that it is distinctively unsuited to address them.

Ultimately, to understand externalities in terms of domination is to set a demanding standard for climate politics—but not a rigid one. It would not permit treating some communities of people and the places where they live as "sacrifice zones," even in the interest of raising living standards for or expanding the capacities of others. At the same time, it would not prohibit any activity outright, even those with harmful byproducts or effects.[135] It would simply allow us to take a more genuine responsibility for these decisions. After all, Friedman and Coase are right that decisions about whether and which pollution is worth its costs should be actively made rather than handed down by economists or moral philosophers. My claim is simply that these decisions should be made collectively by people treating one another as social equals. As Wendy Brown reminds us, freedom is not only paired with but *is* responsibility.[136] When we blame ourselves for the destruction of the world, we act as if we were already free. I do not think that we are.

NOTES

1 As Douglas Kysar also observes in his chapter, "Ways Not to Think About Climate Change," in this volume.

2 Nicholas Stern, *The Economics of Climate Change: The Stern Review* (Cambridge: Cambridge University Press, 2007), 27.

3 For a few representative examples, see Gernot Wagner and Martin Weitzman, *Climate Shock: The Economic Consequences of a Hotter Planet* (Princeton, NJ: Princeton University Press, 2015); William Nordhaus, *The Climate Casino: Risk, Uncertainty, and Economics for a Warming World* (New Haven, CT: Yale University Press, 2013); Gilbert E. Metcalf, *Paying for Pollution: Why a Carbon Tax Is Good for America* (Oxford: Oxford University Press, 2019).

4 See, for example, John Broome, *Climate Matters: Ethics in a Warming World* (New York: W.W. Norton, 2012); Ravi Kanbur and Henry Shue, *Climate Justice: Integrating Economics and Philosophy* (Oxford: Oxford University Press, 2018); Mark Budolfson, Tristam McPherson, and David Plunkett, *Climate Change and Philosophy* (Oxford: Oxford University Press, 2021); John O'Neill, *Ecology, Policy, and Politics: Human Well-Being and the Natural*

World (London: Routledge, 1993); Mark Sagoff, *The Economy of the Earth: Philosophy, Law, and the Environment*, 2nd ed. (Cambridge: Cambridge University Press, 2008); Michael J. Sandel, "It's Immoral to Buy the Right to Pollute," *New York Times*, December 15, 1997; Michael Sandel, *What Money Can't Buy: The Moral Limits of Markets* (New York: Farrar, Straus, and Giroux, 2012); Debra Satz, *Why Some Things Should Not Be for Sale: The Moral Limits of Markets* (Oxford: Oxford University Press, 2010).

5 Melissa Lane notes this surprising absence in a recent review of the literature: see "Political Theory on Climate Change," *Annual Review of Political Science* 19 (2016): 107–123.

6 Steven G. Medema, *The Hesitant Hand: Taming Self-Interest in the History of Economic Ideas* (Princeton, NJ: Princeton University Press, 2009).

7 Jon Elster, *Logic and Society: Contradictions and Possible Worlds* (Chichester and New York: John Wiley & Sons, 1978), 108; Jon Elster, *Making Sense of Marx* (New York: Cambridge University Press, 1984), 24–26. On unintended consequences, see Daniel Luban, "What Is Spontaneous Order?" *American Political Science Review* 14, no. 1 (2020): 68–80; Richard Vernon, "Unintended Consequences," *Political Theory* 7, no. 1 (1979): 57–73.

8 Jean-Paul Sartre, *Critique of Dialectical Reason*, vol. 1, ed. Jonathan Rée, trans. Alan Sheridan-Smith, (London: Verso, 2004); Garrett Hardin, "The Tragedy of the Commons," *Science* 162, no. 3859 (December 1968): 1243–1248; Ulrich Beck, *Risk Society: Towards a New Modernity*, trans. Mark Ritter (London: SAGE, 1992).

9 A. C. Pigou, *The Economics of Welfare* (London: Macmillan, 1920); Nicholas Stern, "The Economics of Climate Change," *American Economic Review* 98, no. 2 (2008): 1–37. See for instance Rachel Carson, *Silent Spring* (Boston: Houghton Mifflin, 1962); E. J. Mishan, *The Costs of Economic Growth* (New York: F. A. Praeger, 1967); E. J. Mishan, "The Postwar Literature on Externalities: An Interpretative Essay," *Journal of Economic Literature* 9, no. 1 (1971): 1–28; Fred Hirsch, *Social Limits to Growth* (Cambridge, MA: Harvard University Press, 1976). See also Medema, *The Hesitant Hand*.

10 Kysar, "Ways Not to Think About Climate Change."

11 Ellen Meiksins Wood, "The Question of Market Dependence," *Journal of Agrarian Change* 2, no. 1 (2002): 50–87; Robert Brenner, "The Agrarian Roots of European Capitalism," *Past & Present* 97 (1982): 16–113.

12 For a few important works in a considerable literature, see Steve Vanderheiden, *Atmospheric Justice: A Political Theory of Climate Change* (Oxford: Oxford University Press, 2008); Stephen M. Gardiner, *A Perfect Moral Storm: The Ethical Tragedy of Climate Change* (Oxford: Oxford University Press, 2011); Henry Shue, *Climate Justice: Vulnerability and Protection* (Oxford: Oxford University Press, 2014); Simon Caney, "Climate Change, Intergenerational Equity and the Social Discount Rate," *Politics, Philosophy &*

Economics 13, no. 4 (2012): 320–342; Lucas Chancel and Thomas Piketty, "Carbon and Inequality: From Kyoto to Paris" (Paris School of Economics, November 2015) http://piketty.pse.ens.fr/les/ChancelPiketty2015.pdf. I do not explicitly address Derek Parfit's "Non-Identity" Problem, which I think has been an unfortunate distraction for many philosophers interested in climate change.

13 Respectively, Gardiner, *A Perfect Moral Storm*, and Jane Bennett, *Vibrant Matter: A Political Ecology of Things* (Durham, NC: Duke University Press, 2010).

14 Derek Parfit, *Reasons and Persons* (Oxford: Oxford University Press, 1984); see discussion in Vanderheiden, *Atmospheric Justice*.

15 Iris Marion Young, *Responsibility for Justice* (Oxford: Oxford University Press, 2011), 11.

16 Most notably in Philip Pettit, *Republicanism: A Theory of Freedom and Government* (Oxford: Oxford University Press, 1997), 205; see also Philip Pettit, "Freedom in the Market," *Politics, Philosophy & Economics* 5, no. 2 (2006): 131–149.

17 For central texts in the burgeoning "radical republican" tradition, see Alex Gourevitch, *From Slavery to the Cooperative Commonwealth* (Cambridge: Cambridge University Press, 2014); Bruno Leipold, Karma Nabulsi, and Stuart White, eds., *Radical Republicanism: Recovering the Tradition's Popular Heritage* (Oxford: Oxford University Press, 2020); Alex Gourevitch and Corey Robin, "Freedom Now," *Polity* 52, no. 3 (2020): 384–398; Tom O'Shea, "Socialist Republicanism," *Political Theory* 48, no. 5 (2020): 548–572; Lillian Cicerchia, "Structural Domination in the Labor Market," *European Journal of Political Theory* 21, no. 1 (2019): 1–21; Elizabeth Anderson, *Private Government* (Princeton, NJ: Princeton University Press, 2017); for an argument about domination applied to environmental questions, see Sharon Krause, "Environmental Domination," *Political Theory* 48, no. 4 (2020): 443–468.

18 See Søren Mau, *Mute Compulsion* (London: Verso, 2023); Max Horkheimer and Theodor W. Adorno, *Dialectic of Enlightenment*, ed. Gunzelin Schmid Noerr, trans. Edmund Jephcott (Stanford, CA: Stanford University Press, 2002); Moishe Postone, *Time, Labor, and Social Domination: A Reinterpretation of Marx's Critical Theory* (Cambridge: Cambridge University Press, 1996), 182; Michael Heinrich, *An Introduction to the Three Volumes of Karl Marx's Capital*, trans. Alex Locascio (New York: Monthly Review Press, 2012); William Clare Roberts, *Marx's Inferno: The Political Theory of Capital* (Princeton, NJ: Princeton University Press, 2017).

19 Eric MacGilvray, *Liberal Freedom: Pluralism, Polarization, and Politics* (Cambridge: Cambridge University Press, 2022).

20 For further discussion of materially mediated social relations, see Alyssa Battistoni, *Free Gifts: Capitalism and the Politics of Nature* (Princeton, NJ: Princeton University Press, 2025).

21 For more politically programmatic statements see Kate Aronoff, Alyssa Battistoni, Daniel Aldana Cohen, and Thea Riofrancos, *A Planet to Win: Why We Need a Green New Deal* (New York: Verso, 2019); Alyssa Battistoni, "Climate Still Changes Everything," *Dissent* (Spring 2023); Alyssa Battistoni, "Sustaining Life on This Planet," in *Democratize Work: The Case for Reorganizing the Economy*, ed. Isabelle Ferreras, Julie Battilana, and Dominique Méda (Chicago: University of Chicago Press, 2022), 103–110.

22 Kysar, "How Not to Think About Climate Change," 39.

23 R. D. Collison Black, Alfred William Coats, and Craufurd Goodwin, eds., *The Marginal Revolution in Economics: Interpretation and Evaluation* (Durham, NC: Duke University Press, 1973).

24 A. C. Pigou, *Wealth and Welfare* (London: Macmillan, 1912), 3; see also Ian Kumekawa, *The First Serious Optimist: A. C. Pigou and the Birth of Welfare Economics* (Princeton, NJ: Princeton University Press, 2017); Medema, *The Hesitant Hand*.

25 Alfred Marshall, *Principles of Economics*, 2nd ed., vol. 1 (London: Macmillan, 1891), 76.

26 Published in its first edition as *Wealth and Welfare* (1912).

27 On Pigou's notion of welfare, see Philipp Lepenies, *The Power of a Single Number: A Political History of GDP* (New York: Columbia University Press, 2016); Kumekawa, *The First Serious Optimist*.

28 Pigou, *Economics of Welfare*, 31.

29 Pigou, *Economics of Welfare*, 160.

30 It is not surprising that a British economist in this time wrote about smoke: Pigou cited the astonishing observation that in London, "owing to the smoke, there is only 12 percent as much sunlight as is astronomically possible, and that one fog in five is directly caused by smoke alone." Pigou, *Economics of Welfare*, 160fn3.

31 Pigou, *Economics of Welfare*, 161.

32 Pigou himself noted that Smith's view of the invisible hand was more nuanced than it was often portrayed. Medema, *The Hesitant Hand*; see also Eric MacGilvray, *The Invention of Market Freedom* (Cambridge: Cambridge University Press, 2011); Albert O. Hirschman, *The Passions and the Interests: Arguments for Capitalism Before Its Triumph* (Princeton, NJ: Princeton University Press, 1977).

33 Pigou, *Economics of Welfare*, 111–113, 493. To this day, economists term such taxes a "Pigovian" approach. In particular, Pigou referenced Smith's argument in the *Wealth of Nations* that government should provide

goods that markets would not. See Adam Smith, *An Inquiry into the Nature and Causes of the Wealth of Nations* (Oxford: Oxford University Press, 1976).

34 Tibor Scitovsky, "Two Concepts of External Economies," *Journal of Political Economy* 62, no. 2 (1954): 143; see also Steven G. Medema, "'Exceptional and Unimportant?': Externalities, Competitive Equilibrium, and the Myth of a Pigovian Tradition," *History of Political Economy* 52, no. 1 (2020): 135–170.

35 Carson, *Silent Spring*; Mishan, *The Costs of Economic Growth*; Hirsch, *Social Limits to Growth*.

36 Medema, *The Hesitant Hand*; Frank Hahn, "Reflections on the Invisible Hand," Warwick Economics Research Paper Series (Coventry: University of Warwick, 1981). On the shift in economics, see Roger E. Backhouse, "Economics," in *The History of the Social Sciences since 1945*, ed. Roger E. Backhouse and Philippe Fontaine (Cambridge: Cambridge University Press, 2010), 38–70; Lionel Robbins, *An Essay on the Nature and Significance of Economic Science*, 1st ed. (London: Macmillan, 1932); Lionel Robbins, "Interpersonal Comparisons of Utility: A Comment," *Economic Journal* 48, no. 192 (December 1938): 635–641; Nicholas Kaldor, "Welfare Propositions of Economics and Interpersonal Comparisons of Utility," *Economic Journal* 49, no. 195 (1939): 549–552; J. R. Hicks, "The Foundations of Welfare Economics," *Economic Journal* 49, no. 196 (1939): 696–712.

37 Friedrich Hayek, *The Road to Serfdom: Texts and Documents*, ed. Bruce Caldwell (Chicago: University of Chicago Press, 2007); Milton Friedman, *Capitalism and Freedom* (Chicago: University of Chicago Press, 1962). See also William Callison and Zak Manfredi, *Mutant Neoliberalism: Market Rule and Political Rupture* (New York: Fordham University Press, 2020); Wendy Brown, *Undoing the Demos: Neoliberalism's Stealth Revolution* (New York: Zone, 2015).

38 See an extensive discussion in Medema, *The Hesitant Hand*, 26–53; on this general phenomenon see also Melissa Lane, *Eco-Republic: What the Ancients Can Teach Us About Ethics, Virtue, and Sustainable Living* (Princeton, NJ: Princeton University Press, 2012), 66–69.

39 Nozick's entire argument for a minimal state would ultimately rely on a complicated account of compensation for "boundary crossing"—in effect, a redescription of the problem of the externality. Robert Nozick, *Anarchy, State, and Utopia* (New York: Basic Books, 1974).

40 Ludwig von Mises, *Liberalism: The Classic Tradition*, trans. Ralph Raico (Indianapolis, IN: Liberty Fund, 2005 [1927]), 30, 32; Hayek, *The Road to Serfdom*, 40.

41 Notably, his critique of government intervention was Friedman's most salient departure from Smith, then being revived as a proto-Chicago School thinker. Milton Friedman, "Adam Smith's Relevance for 1976," in

The Indispensable Milton Friedman: Essays on Politics and Economics, ed. Lanny Ebenstein (Washington, DC: Regnery, 2012), 45–46; on the recovery of Smith see Glory M. Liu, *Adam Smith's America: How a Scottish Philosopher Became an Icon of American Capitalism* (Princeton, NJ: Princeton University Press, 2022).

42 For further discussion of "neighborhood effects," see Friedman, *Capitalism and Freedom*; Milton Friedman, *There's No Such Thing as a Free Lunch* (Chicago: Open Court, 1975); Milton Friedman and Rose D. Friedman, *Free to Choose: A Personal Statement* (New York: Harcourt Brace Jovanovich, 1980).

43 R. H. Coase, "The Problem of Social Cost," *Journal of Law & Economics* 3 (1960): 1–44; see also Steven G. Medema, "Neither Misunderstood nor Ignored: The Early Reception of Coase's Wider Challenge to the Analysis of Externalities," *History of Economic Ideas* 22, no. 1 (2014): 111–132.

44 Coase, "The Problem of Social Cost," 13.
45 Coase, "The Problem of Social Cost," 18, 42.
46 Coase, "The Problem of Social Cost," 34.
47 Coase, "The Problem of Social Cost," 42.
48 Coase, "The Problem of Social Cost," 18, 41–42.
49 Coase, "The Problem of Social Cost," 41.
50 Coase, "The Problem of Social Cost," 88.

51 Assuming no transaction costs—a condition that Coase freely acknowledged was rarely met in practice. See R. H. Coase, "Notes on the Problem of Social Cost," in *The Firm, the Market, and the Law* (Chicago: University of Chicago Press, 1988), 158.

52 For the "Coase Theorem," see George J. Stigler, *The Theory of Price*, 3rd ed. (New York: Macmillan, 1966). On Coase's dislike for the theorem, see "Notes on the Problem of Social Cost"; see also Deirdre McCloskey, "The So-Called Coase Theorem," *Eastern Economic Journal* 24, no. 3 (1998): 367–371; Steven G. Medema, "A Case of Mistaken Identity: George Stigler, 'The Problem of Social Cost,' and the Coase Theorem," *European Journal of Law and Economics* 31, no. 1 (2011): 11–38. On the significance of the Coase Theorem, and externalities more generally, to neoliberalism, see Thomas Biebricher, *The Political Theory of Neoliberalism* (Stanford, CA: Stanford University Press, 2018).

53 See for example Herman E. Daly, *Steady-State Economics: The Economics of Biophysical Equilibrium and Moral Growth* (New York: W. H. Freeman, 1977); Herman E. Daly and Kenneth N. Townsend, eds., *Valuing the Earth: Economics, Ecology, Ethics* (Cambridge, MA: MIT Press, 1993); Nicholas Georgescu-Roegen, *The Entropy Law and the Economic Process* (Cambridge, MA: Harvard University Press, 1979); William D. Nordhaus, "World Dy-

namics: Measurement Without Data," *Economic Journal* 83, no. 332 (1973): 1156–1183; William D. Nordhaus and James Tobin, "Is Growth Obsolete?," *Economic Research: Retrospect and Prospect*, Vol. 5: Economic Growth (New York: National Bureau of Economic Research, 1972); Elinor Ostrom, *Governing the Commons: The Evolution of Institutions for Collective Action* (Cambridge: Cambridge University Press, 1990).

54 Both, then, are generally consistent with what Mark Budolfson has described as a "Default Libertarian" approach, i.e., not "command and control": see Mark Budolfson, "Market Failure, the Tragedy of the Commons, and Default Libertarianism in Contemporary Economics and Policy," in *The Oxford Handbook of Freedom*, ed. David Schmidtz and Carmen E. Pavel (Oxford: Oxford University Press, 2017).

55 For the longer history of these ideas, see MacGilvray, *The Invention of Market Freedom*.

56 John Rawls, for example, discusses instances of "public harms, as when industries sully and erode the natural environment" in explicitly Pigovian terms in *A Theory of Justice* (Cambridge, MA: Harvard University Press, 1971).

57 Michael J. Sandel, "It's Immoral to Buy the Right to Pollute," *New York Times*, December 15, 1997; see also Michael J. Sandel, *What Money Can't Buy: The Moral Limits of Markets* (New York: Farrar, Straus and Giroux, 2012), 73–75; Robert Goodin, "Selling Environmental Indulgences," *Kyklos* 47, no. 4 (1994): 573–596.

58 Wilfred Beckerman and Joanna Pasek, "The Morality of Market Mechanisms to Control Pollution," *World Economics* 4, no. 3 (2003): 191–207; Simon Caney and Cameron Hepburn, "Carbon Trading: Unethical, Unjust and Ineffective?," *Royal Institute of Philosophy Supplement* 69 (2011): 201–234.

59 Karl William Kapp, *The Social Costs of Private Enterprise* (Cambridge, MA: Harvard University Press, 1950); on Kapp, see Sebastian Berger, *The Social Costs of Neoliberalism: Essays on the Economics of K. William Kapp* (Nottingham: Spokesman Books, 2017); Sebastian Berger, "K. William Kapp's Social Theory of Social Costs," *History of Political Economy* 47, no. S1 (2015): 227–252.

60 Kapp, *Social Costs*, 91.

61 Kapp, *Social Costs*, 231–233.

62 Kapp, *Social Costs*, 16; Karl Polanyi, *The Great Transformation: The Political and Economic Origins of Our Time* (Boston: Beacon Press, 2001 [1944]).

63 A similar idea underpins the ecosocialist James O'Connor's "second contradiction" thesis, which argues that environmental movements might impose costs on capital that will lead to economic crisis: see "Capi-

talism, Nature, Socialism: A Theoretical Introduction," *Capitalism Nature Socialism* 1, no. 1 (January 1, 1988): 11–38.

64 Intergovernmental Panel on Climate Change, *Global Warming of 1.5°C*, ed. V. Masson-Delmotte, P. Zhai, H.-O. Pörtner, D. Roberts, J. Skea, P. R. Shukla, A. Pirani, W. Moufouma-Okia, C. Péan, R. Pidcock, S. Connors, J.B.R. Matthews, Y. Chen, X. Zhou, M. I. Gomis, E. Lonnoy, T. Maycock, M. Tignor, and T. Waterfield (Geneva, 2018).

65 Coase, "The Problem of Social Cost," 44.

66 This is why, although Condon is right that climate policy has recently shifted toward programs intended to spur innovation more directly, I don't see this as a refutation of the externality framework so much as a shift in focus and angle of attack.

67 Milton Friedman, "Free Trade and the Steel Industry" (Lecture, Utah State University, 1978).

68 Lawrence H. Summers, "Memo on 'Dirty' Industries" (World Bank, December 12, 1991).

69 For an exemplary critique see David Pellow, *Resisting Global Toxics: Transnational Movements for Environmental Justice* (Cambridge, MA: MIT Press, 2007).

70 Satz, *Why Some Things Should Not Be for Sale*, 95–97.

71 Steven Medema, "The Coase Theorem at Sixty," *Journal of Economic Literature* 58, no. 4 (2020): 1045–1128; Paul A. Samuelson, "Some Uneasiness with the Coase Theorem," *Japan and the World Economy* 7, no. 1 (1995): 1–7; for a particularly creative solution, which I cannot address here, see Christopher Stone, "Should Trees Have Standing?: Toward Legal Rights for Natural Objects," *Southern California Law Review* 45 (1972): 450–501.

72 G. A. Cohen, "Are Workers Who Take Hazardous Jobs Forced to Take Hazardous Jobs?," in *History, Labour, and Freedom: Themes from Marx* (Oxford: Clarendon Press, 1988), 252; for a more basic statement of this argument, see G. A. Cohen, "The Structure of Proletarian Unfreedom," *Philosophy & Public Affairs* 12, no. 1 (1983): 3–33; G. A. Cohen, "Capitalism, Freedom, and the Proletariat," in *On the Currency of Egalitarian Justice, and Other Essays in Political Philosophy* (Princeton, NJ: Princeton University Press, 2011), 147–165.

73 As Hayek puts it, "even if the threat of starvation to me and perhaps to my family impels me to accept a distasteful job at a very low wage . . . I am not coerced by him or by anybody else." Friedrich A. Hayek, *The Constitution of Liberty* (Chicago: University of Chicago Press, 1960), 13.

74 Cohen, "Are Workers Who Take Hazardous Jobs Forced to Take Hazardous Jobs?," 252. See also Nicholas Vrousalis, *Exploitation as Domination: What Makes Capitalism Unjust* (Cambridge: Cambridge University

Press, 2023); A. J. Julius, "The Possibility of Exchange," *Politics, Philosophy & Economics* 12, no. 4 (2013): 361–374.

75 Cf. Anderson, *Private Government*; Steven Klein, "Fictitious Freedom: A Polanyian Critique of the Republican Revival," *American Journal of Political Science* 61, no. 4 (2017): 852–863; Alex Gourevitch, "Labor Republicanism and the Transformation of Work," *Political Theory* 41, no. 4 (2013): 591–617; Cicerchia, "Structural Domination."

76 See especially Klein, "Fictitious Freedom"; Gourevitch, "Labor Republicanism"; Gourevitch, *From Slavery to the Cooperative Commonwealth*; O'Shea, "Socialist Republicanism"; O'Shea, "Radical Republicanism and the Future of Work," *Theory & Event* 24, no. 4 (2021): 1050–1067.

77 Gourevitch, "Labor Republicanism," 596. Pettit's reading of the market, by contrast, explicitly assumes that "imbalances of property and power" do *not* generate domination within relations of exchange. Pettit, "Freedom in the Market," 142.

78 Vrousalis, *Exploitation as Domination*, 1.

79 EPA, "Climate Change and Social Vulnerability in the United States: A Focus on Six Impacts," US Environmental Protection Agency, EPA 2021. 430-R-21–003. www.epa.gov.

80 G. A. Cohen, "Freedom and Money," in *On the Currency of Egalitarian Justice, and Other Essays in Political Philosophy* (Princeton, NJ: Princeton University Press, 2011), 175.

81 See discussion in George Caffentzis, "The Work/Energy Crisis and the Apocalypse," *Midnight Notes* no. 3 (1980): 26.

82 See Donald MacKenzie, "Constructing Emissions Markets," in *Material Markets: How Economic Agents Are Constructed* (Oxford: Oxford University Press, 2009), 137–176.

83 Cohen, "Are Workers Who Take Hazardous Jobs Forced to Take Hazardous Jobs?"

84 Simon Caney, "Climate Change, Energy Rights, and Equality," in *The Ethics of Global Climate Change*, ed. Denis G. Arnold (Cambridge: Cambridge University Press, 2011), 77–103; for an argument for the inalienable right to a healthy environment see Henry Shue, "Bequeathing Hazards: Security Rights and Property Rights of Future Humans," in *Global Environmental Economics*, ed. Mohammed H. I. Dore and Timothy D. Mount (Malden, MA: Blackwell, 1999), 40–42.

85 See Joan Martinez-Alier, "From Political Economy to Political Ecology," in *Bioeconomics and Sustainability: Essays in Honor of Nicholas Georgescu-Roegen*, ed. K. Mayumi and J. M. Gowdy (Cheltenham UK: Edward Elgar 1999), 39.

86 See Pellow, *Resisting Global Toxics*; see also Sara Holiday Nelson, "Neoliberal Environments: Crisis, Counterrevolution, and the Nature

of Value" (PhD diss., University of Minnesota, 2017); Michael Denning, "Wageless Life," *New Left Review*, no. 66 (December 1, 2010): 79–97.

87 On articulation see Stuart Hall, "Race, Articulation, and Societies Structured in Dominance [1980]," in *Essential Essays, Vol. 1*, ed. David Morley (Durham, NC: Duke University Press, 2018), drawing on the notion of "articulation" in Louis Althusser, *For Marx* (London: Allan Lane, 1965).

88 Laura Pulido, "Flint, Environmental Racism, and Racial Capitalism," *Capitalism Nature Socialism* 27, no. 3 (2016): 1–16; Laura Pulido, "Geographies and Race and Ethnicity II: Environmental Racism and Racial Capitalism," *Progress in Human Geography* 41, no. 4 (2016): 524–533.

89 For a few key works in a huge literature see David N. Pellow and Robert J. Brulle, *Power, Justice, and the Environment: A Critical Appraisal of the Environmental Justice Movement* (Cambridge, MA: MIT Press, 2005); Liam Downey and Brian Hawkins, "Race, Income, and Environmental Inequality in the United States," *Sociological Perspectives* 51, no. 4 (2008): 759–781; Charles W. Mills, "Black Trash," in *Faces of Environmental Racism: Confronting Issues of Global Justice*, ed. Laura Westra and Bill E. Lawson (New York: Rowman & Littlefield, 2001), 73–93. On race and residential patterns, see Dorceta Taylor, *Toxic Communities: Environmental Racism, Industrial Pollution, and Residential Mobility* (New York: New York University Press, 2014).

90 Laura Pulido, *Environmentalism and Economic Justice: Two Chicano Struggles in the Southwest* (Tucson: University of Arizona Press, 1996).

91 Olúfẹ́mi O. Táíwò, *Reconsidering Reparations* (Oxford: Oxford University Press, 2022), 162–166.

92 Thomas Wiedmann, Manfred Lenzen, Lorenz T. Keyßer, and Julia K. Steinberger, "Scientists' Warning on Affluence," *Nature Communications* 11, no. 1 (2020): 3107.

93 Naomi Oreskes and Erik Conway, *Merchants of Doubt: How a Handful of Scientists Obscured the Truth on Issues from Tobacco Smoke to Global Warming* (New York: Bloomsbury Press, 2010); Henry Shue, "Reckless Complicity: International Banks and Future Climate," in *The Routledge Handbook of Philosophy of Responsibility*, ed. Maximilian Kiener (London: Routledge, 2024), 431–441; Leah Cardamore Stokes, *Short-Circuiting Policy: Interest Groups and the Battle Over Clean Energy and Climate Policy in the American States* (Oxford: Oxford University Press, 2020).

94 See, for example, Elizabeth Cripps, *Climate Change and the Moral Agent: Individual Duties in an Interdependent World* (Oxford: Oxford University Press, 2013); Vanderheiden, *Atmospheric Justice*; Julia Nefsky, "Collective Harm and the Inefficacy Problem," *Philosophy Compass* 14, no. 4 (2019): e12587; Dale Jamieson, "Climate Change, Responsibility and Justice," *Science and Engineering Ethics* 16, (2010): 431–445; Dale Jamieson,

"Ethics, Public Policy and Global Warming," *Science, Technology, and Human Values* 17 (1992): 139–153; Stephen Gardiner, "Is No One Responsible for Global Environmental Tragedy? Climate Change as a Challenge to Our Ethical Concepts," in *The Ethics of Global Climate Change*; Douglas MacLean, "Climate Complicity and Individual Accountability," *The Monist* 102, no. 1 (2019): 1–21; Elisabeth Anker, *Ugly Freedoms* (Durham, NC: Duke University Press, 2021).

95 Judith Lichtenberg, "Negative Duties, Positive Duties, and the 'New Harms'," *Ethics* 120, no. 3 (2010): 557–578.

96 Lichtenberg, "Negative Duties," 558.

97 Parfit, *Reasons and Persons*; see discussion in Vanderheiden, *Atmospheric Justice*. For arguments rooted in similar reasoning see Broome, *Climate Matters*; John Broome, "Against Denialism," *The Monist* 102, no. 1 (2019): 110–129.

98 Lichtenberg, "Negative Duties," 560.

99 Cited in Leif Wenar, *Blood Oil: Tyrants, Violence, and the Rules that Run the World* (Oxford: Oxford University Press, 2017). For a related but different analysis of moral consumerism see Tad Skotnicki, *The Sympathetic Consumer: Moral Critique in Capitalist Culture* (Stanford, CA: Stanford University Press, 2021).

100 Wenar, *Blood Oil*, xx.

101 See Friedrich A. Hayek, "The Use of Knowledge in Society," *American Economic Review* 35, no. 4 (1945): 519–530.

102 MacGilvray, *Liberal Freedom*, 119

103 MacGilvray, *Liberal Freedom*, 103. Italics in original.

104 See Luban, "What Is Spontaneous Order?"; Medema, *The Hesitant Hand*.

105 Pettit, *Republicanism*, 205; Pettit, "Freedom in the Market."

106 Pettit, "Freedom in the Market," 135; for a critique see Steven Klein, "Fictitious Freedom: A Polanyian Critique of the Republican Revival," *American Journal of Political Science* 61, no. 4 (2017): 852–863; Roberts, *Marx's Inferno*, 85n. It might nevertheless be possible to hold market participants responsible, in part, for the outcomes that they know markets are likely to produce, if they have alternatives to participating in them. But in capitalist societies, people generally do *not* have reasonable alternatives to market participation, all the more so if they do not themselves own the elements necessary for self-sufficiency (e.g., land, tools, etc.). Thanks to Chiara Cordelli for pressing this point.

107 Luban, "What Is Spontaneous Order?"

108 Hayek, *Constitution of Liberty*, 137; Hayek, *Road to Serfdom*, 216–217; see discussion in MacGilvray, *The Invention of Market Freedom*, 172; Gourevitch and Robin, "Freedom Now."

109 Hayek, *Constitution of Liberty*, 143.
110 Jan Kandiyali, "Should Socialists Be Republicans?" *Critical Review of International Social and Political Philosophy* 27, no. 7 (2024): 1032–1049.
111 Roberts, *Marx's Inferno*, 85n; William Clare Roberts, "Reading Capital as Political Theory: On the Political Theory of the Value-Form," in *Marx's Capital after 150 Years: Critique and Alternative to Capitalism*, ed. Marcello Musto (London: Routledge, 2019), 228.
112 Hayek, *Road to Serfdom*, 210.
113 See Daniel Luban, "In Marx's Republic," *The Nation*.
114 Sartre, *Critique*, 253–342; see also discussions in Young, *Responsibility for Justice*, 155–157. For a more extensive discussion of seriality see Battistoni, *Free Gifts*.
115 As discussed in Sean Irving, *Hayek's Market Republicanism: The Limits of Liberty* (New York: Routledge, 2020).
116 Hayek, *Constitution of Liberty*, 137; Hayek, *Road to Serfdom*, 216–217; see discussion in MacGilvray, *The Invention of Market Freedom*, 172; Gourevitch and Robin, "Freedom Now."
117 Hayek, *Road to Serfdom*, 216–217; see discussion in Gourevitch and Robin, "Freedom Now."
118 Hayek, *Constitution of Liberty*, 134.
119 Jean-Paul Sartre, *Being and Nothingness: An Essay in Phenomenological Ontology*, trans. Sarah Richmond (London: Washington Square Press, 2018 [1943]), 78; see also Sartre, *Existentialism Is a Humanism*, trans. Carolyn Macomber (New Haven, CT: Yale University Press, 2007 [1946]); Simone de Beauvoir, *Philosophical Writings*, ed. Margaret A. Simons (Urbana: University of Illinois Press, 2004); Beauvoir, *The Ethics of Ambiguity*, trans. Bernard Frechtman (New York: Open Road, 2018 [1947]).
120 For further discussion see Battistoni, *Free Gifts*.
121 Sartre, *Critique*, 281–282.
122 There are, of course, exceptions, but economics assumes that people are price-sensitive, and I think rightly so. If this weren't true, then market solutions wouldn't work and we needn't bother discussing them at all.
123 William Clare Roberts calls this *akrasia*, the failure to act on one's own judgment. Roberts, *Marx's Inferno*, 56–70.
124 *Pace* Theodor Adorno, *Minima Moralia: Reflections on a Damaged Life* (New York: Verso, 2005).
125 See Pellow, *Resisting Global Toxics*; Pulido, "Geographies of Race and Ethnicity"; on the "race to the bottom" see Gourevitch, "Labor Republicanism"; Benjamin McKean, *Disorienting Neoliberalism: Global Justice and the Outer Limit of Freedom* (Oxford: Oxford University Press, 2020).
126 Hayek, *Road to Serfdom*, 211.

127 See Iris Marion Young, *Responsibility for Justice* (Oxford: Oxford University Press, 2011).

128 Cf. Luban, "What Is Spontaneous Order?"

129 Brad Plumer, "New U.N. Climate Report Says Put a High Price on Carbon," *New York Times*, October 8, 2018; OECD, "Few Countries Are Pricing Carbon High Enough to Meet Climate Targets," www.oecd.org.

130 For a discussion see Alyssa Battistoni and Geoff Mann, "Climate Bidenomics," *New Left Review* 143 (September/October 2023): 55–77.

131 Matto Mildenberger, *Carbon Captured: How Business and Labor Control Carbon Politics* (Cambridge, MA: MIT Press, 2020), 236.

132 Contra Budolfson's suggestion that weather itself may be oppressive, I do not think that nature *itself* dominates human beings; nor do I think that nature can be dominated *by* human beings, as Sharon Krause has recently argued. See Krause, "Environmental Domination"; William Leiss, *The Domination of Nature* (New York: George Braziller, 1972).

133 Krause, "Environmental Domination."

134 Iris Marion Young, *Justice and the Politics of Difference* (Princeton, NJ: Princeton University Press, 1990), 38.

135 For this reason I don't think this analysis is vulnerable to the thought experiment Budolfson presents, concerning the residents of another newly discovered planet who beg Earthlings for access to coal.

136 Wendy Brown, *Edgework: Critical Essays on Knowledge and Politics* (Princeton, NJ: Princeton University Press, 2005).

4

ENVIRONMENTAL JUSTICE, CAPITALISM, DEMOCRACY

MARK BUDOLFSON

How Standard Economic Environmental Policy Ignores Inequalities

One of the most problematic aspects of standard economic policy responses to negative externalities (such as air pollution and greenhouse gas emissions) is that these policies tend to ignore the socioeconomic distribution of both the externality itself and the distribution of the costs and benefits of the policy response. Therefore, these policies often fail to redress important inequalities, and can even perpetuate inequality. And because the distribution of the harms from a negative externality is often the most important social problem connected to that externality, this means that by ignoring distributional effects, standard economic policies are at least often ignoring what is most important—and in some cases could be making it worse.[1]

Why are distributional facts ignored by standard economic policy responses to negative externalities? This is a consequence of the fact that standard economic analyses—i.e., the actual policies that tend to be enacted based on the advice of environmental economists—aim to minimize the total dollar cost to the economy of any desired level of negative externality reduction by creating a *uniform* disincentive for the negative externality for all people in all locations. What I mean by a *"uniform" disincentive* is that the very same dollar magnitude of disincentive is in place everywhere: such as a *uniform* carbon tax in which the very same tax is charged everywhere, or a *uniform* cap and trade system in which the very same price for permits is charged everywhere as a consequence of a well-functioning market for the permits. This uniformity in disincentive is a key part of standard economic policy, as it is necessary to ensure that any level of reductions comes at the least cost to the economy.[2]

One problem with this kind of standard economic policy is that it ignores the distribution of the burden of the tax on different socioeconomic groups. Most obviously, it ignores the fact that a uniform tax often disproportionately burdens the poor, since a dollar of tax subtracts more wellbeing from poor households than a dollar of tax to rich households. Standard economic policy ignores these differences in wellbeing impacts from a uniform tax. Beyond this, in for example the case of climate change, the regressive nature of a standard carbon tax in a rich country like the United States is a widely discussed problem, as poor households in richer nations spend a higher fraction of their income on carbon-intensive goods than rich households, making such a tax regressive even beyond accounting for the fact that each dollar spent by a poor household reduces their wellbeing more than a dollar spent by a rich household.[3]

Economists often attempt to justify standard economic policy advice by saying that it is part of their job to intentionally ignore equity, on the grounds that their job is to focus on efficiency, not equity. But at best this would provide a justification for *economists* to ignore equity, and would not provide any justification for *policies* to ignore equity. And in fact the economists who make this argument agree, as they explicitly endorse the idea that equity needs to be taken into account by someone else in policymaking—just not by economists. With that in mind, this line of argument from economists faces a further problem insofar as economists have carefully maneuvered to take the reins of policymaking all for themselves, as they have arguably done in connection with US federal environmental policy as well as many other areas. It can be argued that they have done this because to the extent that they alone control the reins, it is known that their standard economic policies that ignore equity will be enacted in a context in which it is known that no one else will take equity adequately into account in any supporting policies—which means that the policies they end up recommending are ethically unacceptable by their own lights. Furthermore, in such a context, standard economic policy can make outcomes worse rather than better for both individuals and society in the aggregate, as explained earlier, and has been shown in economic theory from a utilitarian perspective.[4]

To illustrate a different and less obvious reason why a uniform disincentive structure is problematic and ignores distributional consequences, consider a simplistic illustration of how such a policy could go wrong: Suppose that a society wants to cut emissions of an air pollutant in half (for example, GHG emissions), and so it ratchets up a uniform tax on that pollutant until emissions are cut in half across society. But, unfortunately, the way that this plays out is there are initially two emission sources—one source in the urban area where most people live, and one larger source in the middle of nowhere—and it turns out that the least-cost way for the economy to deal with the new price on emissions is to close the rural emitting facility altogether, invest in upgrading the controls at the urban emitting facility, and move all the production to the urban plant, which then ends up doing all of the production for society. The result is that pollution actually increases at the urban location where most people live and decreases a lot more in the middle of nowhere. This illustrates how a uniform disincentive scheme across society ignores the distribution of the emissions, and as a result can go wrong by ignoring distributional effects. In this dramatic example, we could imagine that the policy actually makes the outcome worse for society by redistributing a smaller amount of emissions in a way that actually makes the aggregate health burden worse than it was with a higher level of emissions.

And even if we assume there isn't such a dramatic failure of the policy, and thus even if we imagine that health is improved for most people in society by the policy and is only made worse for a few who live very close to the single higher emitting plant that emerges from the policy, it is easy to see that this is itself a problematic distributional outcome that the policy ignores. This is especially problematic if the small group that suffers greater harm consists of those at the bottom of the socioeconomic distribution or the most oppressed racial minority groups.[5]

This last worry is a key reason why environmental justice (EJ) advocates often decry standard economic policy responses to pollutants, as they fear that something very much like this result is the predictable outcome of standard economic environmental policy.

Furthermore, any missed opportunity to alleviate inequalities in pollution exposure is especially important, given that inequalities in exposure to pollutants are one primary cause of structural injustices in society and the intergenerational transmission of inequalities.[6] Specifically, the EJ worry is that the uniform incentives created by such policies will incentivize society's emissions to be located at the places where they can be sited at the least cost, thereby predictably concentrating pollution in the poorest and most vulnerable communities.[7] In this way, the EJ worry is that implementing standard environmental economic policy becomes a way of predictably implementing and deepening structural injustice. (It is a further question to what extent these specific EJ concerns are predictably the consequence of standard environmental economic policy, which I set aside here.)[8]

In sum, standard economic policy ignores many inequalities, and according to some EJ advocates can actually amplify some of the deepest structural injustices in society.

The Problem with Familiar Critiques of Capitalism

All that being said, the observations in the previous section about the inadequacy of standard market-based economic policy are often turned too quickly into arguments against capitalism itself. These familiar but too-quick arguments often have something like the form: "Climate change shows that capitalism is deeply and necessarily flawed, because capitalism is the fundamental underlying problem behind the catastrophic problems associated with climate change and capitalism is also inherently incompatible with any decent solutions to these problems, much less the best solutions to these problems."[9]

The main point of this section is to suggest that these familiar critiques of capitalism are mistaken: In fact, the best solutions to climate change are perfectly consistent with capitalism, and actually require capitalism. So, it is a mistake to argue that climate change shows that there is a deep problem for capitalism: While it is true that we live in (somewhat) capitalist societies, and while it is true that capitalism is not currently well-regulated, and that these two things together are in some sense causing problems such as climate change and other problems of over-pollution, it is incorrect to think that capitalism is the fundamental underlying problem, because in fact well-regulated capitalism provides the best solution to the problem.

To see the intuition behind this, it may be useful to consider the example of air pollution, which has been one of the most harmful externalities of the previous century associated with capitalism. Air pollution was not well-regulated in many places in the past one hundred years (and is still not fully well-regulated even in rich nations today), and the historic number of deaths caused by air pollution have thus far been even greater than the historic deaths resulting from climate change.[10] (Which is not to take a position on what the future holds.) In an earlier era of environmentalism in the late 1960s, it was also common to offer structurally analogous arguments that air pollution revealed a deep problem with capitalism. However, it is now generally agreed that this particular anti-capitalist argument was mistaken, because in fact the best solutions to the problem of air pollution are perfectly consistent with capitalism, and actually require capitalism, e.g., via market-based pricing of emissions.[11]

Generalizing, the fundamental argument here regarding capitalism and externalities is straightforward:

> Premise 1. The best response to externalities (including climate change, air pollution, etc.) is perfectly consistent with capitalism and in fact depends on market mechanisms that are most naturally part of a capitalist system: namely, a Pigouvian tax (e.g., a carbon tax) or other market-based externality pricing system, together within linked policies to use the revenues from the tax on the externality to promote equity, or other straightforward supporting policies that promote justice, such as equity weighting to give more value in the pricing scheme to reductions paid by those who are richer, or other distributional weights that give preference to reductions in places that are overburdened. Indeed, this ideally best response is most naturally achieved within a capitalist system.
> Premise 2. In addition, almost everyone would be much worse off under any alternative response that got rid of capitalism.[12]

> Conclusion. Therefore, not only is capitalism consistent with the best response to externalities such as climate change, it is also a necessary part of any desirable response, and a necessary part of any desirable basic structure for society.

For our purposes here, we can set aside the claim made in the second premise that almost everyone would be much worse off without capitalism. Given that our focus is on the more narrow issue that is the focus of the first premise, the key point here is that many influential experts on environmental policy in recent decades (and most major environmental regulations at the federal level from both parties in the past fifty years) endorse the truth of this first premise.[13]

In sum, the best responses to climate change, air pollution, and other externalities known to policymakers involve pricing the externality (e.g., via a Pigouvian tax), using the revenues to promote equity, or a number of other straightforward supporting policies that promote equity and justice. This suggests that the best solutions to climate change are perfectly consistent with capitalism, and actually require capitalism. Two further supporting data points are that richer capitalist societies have shown the most success in regulating negative externalities once they become sufficiently rich,[14] and noncapitalist nations such as the Soviet Union and North Korea fared dramatically worse than capitalist counterparts with respect to environmental pollution, even controlling for per capita wealth.[15]

At a theoretical level, this suggests a different diagnosis of the root problem behind externalities. The problem is not ultimately with capitalism, but instead with the failure to ensure that capitalism is well-regulated. And in contemporary free societies, the root of this problem is our particular forms of democracy, rather than capitalism. Some data in support of this are that when we consider more authoritarian yet (somewhat) capitalist nations such as China, they are able to move more quickly and efficiently toward regulation of externalities than democratic nations have done, especially when controlling for per capita wealth—as illustrated in the past decade by China's rapid air quality improvements, its movement toward market-based policies such as its National Emissions Trading Scheme, and market-based regulation of other externalities.[16]

Consistent with this, an additional question outside of our focus is whether we should have a positive or negative overall evaluation of China's environmental policies. The point that is important for our purposes is independent of this evaluation. For example, some experts argue that while China is able to move more quickly and efficiently, China nonetheless has worse environmental policies overall because of other problems and distortions caused by China's particular forms of authoritarianism, which ensure that environmental and other "social welfare" policies in China are primarily designed to advance authoritarianism, rather than vice versa, typically at greatest expense to the social welfare of those in China with the lowest socioeconomic status.[17]

Setting these issues aside, for our purposes the important point is that our particular forms of democracy involve important impediments to proper regulation, and these political failings are really the root cause of problems with externalities, rather than capitalism. Perhaps the solution is, as some have argued, that we need "10% less democracy," perhaps in the form of something akin to more insulated institutions such as the Federal Reserve, to govern externalities, thereby insulating regulation of externalities from democratic forces, much as monetary policy has been insulated by the Federal Reserve itself.[18] Or perhaps there is some other even better way to modify our decision-making procedures to yield regulation of externalities.[19] The important point for our purposes is that regulatory failures are not ultimately caused by capitalism; they are caused by failures within particular forms of democracy and other political systems. So, insofar as the root cause of a fundamental underlying problem behind climate change must be located in one of the main mechanisms for social coordination discussed by political theorists, the best place to locate it is in specific forms of democracy, specific forms of authoritarianism, and the like, rather than capitalism. Again, the important point is that the best solutions to climate change are perfectly consistent with capitalism, and actually require capitalism.

BATTISTONI'S CRITIQUE OF CAPITALISM, AND OF STANDARD ECONOMIC ENVIRONMENTAL POLICY

Moving beyond the familiar arguments about capitalism in the last section, Alyssa Battistoni offers a deeper and novel critique of the actual practice of policymaking within capitalist societies. Battistoni correctly identifies and makes conceptually clear several ways that well-intentioned standard economic policy can fail from the perspective of justice even when perfectly implemented. In so doing, Battistoni's critique deepens the critique from the first section above, and makes visible several ways that oppression can remain even when standard economic policy is perfectly implemented.[20] Nevertheless, I shall argue that the take-home message for her critique ultimately should be the same as for the line of critique identified in the first section: Namely, that there are no fundamental problems with capitalism per se, but rather with the actual practice of democratic environmental policymaking—including standard economic policy—which does not typically include any of the straightforward supporting policies described in the second section that are necessary to ensure equity and justice, and instead only includes policy mechanisms that ignore justice and the distribution of benefits and harms (such as a uniform Pigouvian tax without any supporting policies).

Battistoni's novel critique begins by noting that even if we have the sort of environmental policy recommended by economists, the inequalities that remain will ensure the poor and socioeconomically vulnerable remain wrongfully dominated (and harmed, oppressed, and disproportionately burdened) by the negative externality itself. So, standard economic environmental policy, even if fully implemented, would not remove many of the most serious ethical problems with the situation, given that significant amounts of pollution would remain, it would be regressively distributed, and this would involve wrongful domination, harm, and oppression. This critique is very powerful, especially given the insightful way that Battistoni grounds the argument in the detailed analysis outlined above of the ways that standard economic policy ignores inequality. And the novelty of Battistoni's critique ensures it is a valuable contribution to thinking about these issues.

With the critique in hand, Battistoni also suggests that anti-capitalist and anti-consumerism conclusions follow from these legitimate criticisms of standard economic policy. In the remainder of this chapter, I want to highlight a few ways that some aspects of Battistoni's arguments for these further conclusions involve complications that are important and not commonly enough addressed in many discussions in social and political philosophy. Because of their interest, and because they introduce important limits on even subtle critiques of capitalism like Battistoni's, I will focus on making them clear in the rest of this discussion. To that end, we can begin by making explicit three further claims that Battistoni presents in continuing the line of critique above:

A. Capitalism inevitably produces wrongful domination, given that (as we've seen above) domination remains even under fully implemented standard environmental economic policy.
B. The rich within capitalist societies are morally responsible for this domination.
C. Individuals within capitalist society also bear moral responsibility for this domination (although individuals should often focus on change at the structural rather than the individual level).

I will try to quickly highlight what I think are the most difficult complications for each of these three further claims in turn. Doing so will introduce limitations on the force of Battistoni's ultimate critique of capitalism and consumers within capitalist systems.

Regarding claim A: One complication is that even if we agree that standard economic policy is insufficient to remove wrongful domination from capitalism, it doesn't follow that there is no *other* way for capitalism to remove wrongful domination, and thus it doesn't follow that capitalism *inevitably must* produce wrongful domination. In fact, many leading theorists in economics, policy, and law have argued for additional market-friendly methods of going beyond standard economic policy to address inequalities head-on rather than ignore them,[21] including some proposals to improve environmental economic policy to address inequalities head-on and avoid exactly the kind of problems outlined above.[22]

In the absence of an argument that the methods proposed by these experts are doomed to fail, we should consider the possibility that *well-regulated capitalism* in its best form, including incorporating these further market-friendly policy measures, might be able to avoid producing wrongful domination even if current instances of capitalism do not.[23]

Regarding claim B: One complication in the argument for B is that it seems to assume both (i) that capitalism or the rich within capitalist societies are causally responsible for the fact that the poor are poor, and (ii) that the poor have not received benefits from the rich and from capitalism that more than compensate them for the downsides of inequality, their constrained labor option set, and other important limitations on their lives. In highlighting the reliance on (i) and (ii), I do not mean to cast doubt on the idea that these assumptions sometimes hold true in specific cases. Instead, my goal is to highlight that there are important complications due to the fact that the global poor benefit so much (according to some experts) from a globalized capitalist economy, and the wealth of nations arguably has numerous complex causes, some of which arguably do not involve the rich and capitalism making poorer countries poorer. Depending on how far one believes it is empirically plausible to go with these thoughts, one can anticipate those who believe that (at least well-regulated) capitalism would benefit the poor to an extent that more than outweighs the downsides of inequality, constrained labor option sets, etc., and that this fact gets capitalism off the hook for some of the bad aspects of inequality.

As a related conceptual point, from the fact that A knowingly caused x, and that x is badly oppressive to B, it does not follow that A is wrongly oppressing B, and doesn't even obviously follow that A is oppressing B at all. To illustrate, imagine a fanciful (and highly counterfactual) case where a new planet appears to us, which we can see is filled with people in desperate poverty who cannot benefit from direct charity—instead, the only thing we can do to help them is to give them coal to make cheap power for themselves, because a key source of their poverty is a lack of energy. Suppose they see the coal we give them as like manna from heaven, but manna that also causes the presence of oppressive and harmful dust in the air. If we assume for the sake of argument that there is really no other way we can help them, and that they really, really want the coal we give them because it helps them dramatically on balance (they are no longer starving and in darkness), then it seems we are not wrongly oppressing or dominating them by sending them the coal, even if we are thereby causing the dust cloud that is badly oppressive to them. And it is not clear that *we* are oppressing or dominating them at all.

As a side note, even in conditions of perfect justice there could be lots of domination and oppression by nature—the weather can be oppressive after all. As intuition for this, consider those who were born at an earlier time before anything like social safety nets or even large societies existed, who were born into places where the weather frequently destroyed all of their work and repeatedly set them back to square one. For them, it is easy to imagine the weather as an oppressive force dominating their lives—indeed, their thoughts and rituals might themselves be dominated by responding to the domination of the weather. More generally, there are ways in which natural scarcity and other circumstances might be the entity that is "responsible" for the oppression that people sometimes experience, including perhaps the inhabitants of the imaginary poor planet in the example above (if the details of the case are developed in that way).

All of this is especially important to the evaluation of capitalism, if capitalism is analogous to our best biomedical knowledge and treatments: Just as the regrettable and oppressive fact that humans will always be constrained by finite lives pairs naturally with the thought that there are strong reasons to lengthen our lives and health, so too the regrettable and oppressive fact of scarcity pairs naturally with the thought that there are strong reasons to produce more abundance for all—and some would argue that well-regulated capitalism is the best technology we currently have for doing that. For those who endorse this perspective, it is easier to see why unfortunate side effects such as inequalities (especially if well-managed under well-regulated capitalism) could be considered a regrettable side effect that can be nonetheless ethically justified by the larger and more ethically important benefits of (well-regulated) capitalism.

Regarding claim C: From the fact that every individual is a contributor to climate change harms (in whatever form of contribution one likes—causal contribution, consequentialist difference-maker to harm, etc.),[24] it doesn't follow that each is morally responsible for a share of those harms. To illustrate, imagine that zero emissions power is invented soon that is also too cheap to meter, but that one hundred years from now this power technology still hasn't been deployed because of an evil conspiracy by a few leaders. Now consider a child born one hundred years from now, and consider the evaluation that they are responsible for the harms from climate change because all of their power consumption still contributes to climate change. Even if we agree with the premise about their contribution of their consumption, it doesn't follow that we should hold this future person in these circumstances morally responsible for the climate harms that happen—instead, intuitively it is the leaders of the conspiracy who are responsible. Of course our actual situation is very different, but a reasonable worry still remains that a realistic-but-limited analog of these complications remains. For example, if you are disposed to adjust the climate policy lever in the right way if only you could (e.g., with a carbon tax and complementary policies) to make the outcome much better and in a way that is costly to you, and you take good action as a citizen politically and interpersonally in supportive ways, and act well professionally in supportive ways, then this might importantly reduce your individual responsibility for bad outcomes. More generally, there are reasons to be suspicious of the common assumption in ethics and political theory that if you are part of a group that is collectively doing harm, then it follows directly that you are morally responsible for (your proportional share) of the harm. This line of thought may be too quick, and ethical reality may be more complicated.

In sum, Battistoni correctly highlights one of the most problematic aspects of standard economic policy responses to negative externalities such as air pollution and greenhouse gas emissions: namely, that these policies tend to ignore the socioeconomic distribution of both the externality itself and the distribution of the costs and benefits of the policy response. Therefore, these policies often fail to redress important inequalities, and can even perpetuate inequality, domination, and injustice. Battistoni's contribution is particularly important and novel, in part because it makes visible and more conceptually clear several ways that well-intentioned standard economic policy can fail deeply from the perspective of justice even when perfectly implemented by the standards of standard economic policy, which have not been recognized or widely discussed. This is important, because standard economic policy is the dominant approach to responding to negative externalities in many places.[25]

However, the conclusions we should draw from these facts may be more limited than they initially appear. In particular, as we've seen in this section, there are important further complications regarding moral responsibility in the evaluation of capitalism and consumers within such a system. And as we saw in the second section above, none of these observations imply any deep problem for capitalism, because the best solution to problems of externalities are perfectly consistent with capitalism, and actually require capitalism. The problem is not ultimately with capitalism, but instead with the failure to ensure that capitalism is well-regulated.

Acknowledgments

Mark Budolfson was supported by National Science Foundation grant #2420344. Thanks to Melissa Lane, Chiara Cordelli, and participants at the American Society for Political and Legal Philosophy Annual Conference in 2023 at Princeton University for helpful comments and discussions.

Notes

1 See also Alyssa Battistoni, "Domination in the Age of the Externality," in this volume; for general discussion of standard economic policy, distributional complications, and considerations of justice, without Battistoni's special emphasis on structural injustice or environmental injustice, see Daniel Hausman, Michael McPherson, and Debra Satz, *Economic Analysis, Moral Philosophy, and Public Policy*, 3rd ed. (Cambridge: Cambridge University Press, 2016).

2 See for example Lynne Lewis and Tom Tietenberg, *Environmental and Natural Resource Economics*, 12th ed. (New York: Routledge, 2018).

3 See for example Francis Dennig, Mark B. Budolfson, Marc Fleurbaey, Asher Siebert, and Robert H. Socolow, "Inequality, Climate Impacts on the Future Poor, and Carbon Prices," *Proceedings of the National Academy of Sciences* 112, no. 52 (2015): 15827–15832.

4 Graciela Chichilnisky, Geoffrey Heal, and David Starrett, "Equity and Efficiency in Environmental Markets: Global Trade in CO_2 Emis-

sions," in *Environmental Markets: Equity and Efficiency*, ed. Graciela Chichilnisky and Geoffrey Heal (New York: Columbia University Press, 2000).

5 For further discussion along several important dimensions, see for example Nicholas Muller and Robert Mendelsohn, *Using Marginal Damages in Environmental Policy: A Study of Air Pollution in the United States* (Washington, DC: AEI Press, 2013) and Chichilnisky and Heal, eds., *Environmental Markets: Equity and Efficiency*.

6 See for example Anna Aizer and Janet Currie, "The Intergenerational Transmission of Inequality: Maternal Disadvantage and Health at Birth," *Science* 344, no. 6186 (2014): 856–861.

7 See for example Nicky Sheats, "Achieving Emissions Reductions for Environmental Justice Communities Through Climate Change Mitigation Policy," *William and Mary Environmental Law & Policy Review* 41, no. 2 (2017): 377–402.

8 See for example Danae Hernandez-Cortes and Kyle Meng, "Do Environmental Markets Cause Environmental Injustice? Evidence from California's Carbon Market," *Journal of Public Economics* 217 (2023): 104786; Muller and Mendelsohn, *Using Marginal Damages in Environmental Policy*.

9 See Douglas Kysar, "Ways Not to Think About Climate Change," in this volume, and Alyssa Battistoni, "Domination in the Age of the Externality," in this volume, among many others in contemporary discourse more generally who could be cited. For what could be argued is a critique of liberalism similar to this critique of capitalism, see Dale Jamieson and Marcello Di Paola, "Climate Change, Liberalism, and the Public/Private Distinction," in *Philosophy and Climate Change*, ed. Mark Budolfson, Tristram McPherson, and David Plunkett (Oxford: Oxford University Press, 2021), 370–396.

10 See "Death Rate from Air Pollution, World," *Our World in Data*, accessed July 30, 2024, https://ourworldindata.org. For more general discussion of the health and wellbeing impacts of air pollution and other large-scale externalities, see Hannah Ritchie, *Not the End of the World* (New York: Little Brown, 2024).

11 See for example the sort of arguments presented in Robert Stavins, ed., *Economics of the Environment: Selected Readings*, 7th ed. (Cheltenham, UK: Edward Elgar, 2019).

12 For more discussion of the structure of arguments in favor of well-regulated capitalism, based in part on its importance to human wellbeing, see for example Mark Budolfson, "Arguments for Well-Regulated Capitalism, and Implications for Global Ethics, Food, Environment, Climate Change, and Beyond," *Ethics & International Affairs* 35, no. 1 (2021): 83–98, which also discusses other examples of environmental externalities.

13 Stavins ed., *Economics of the Environment.* See also Cass Sunstein, *The Cost-Benefit Revolution* (Cambridge, MA: MIT Press, 2018).

14 See for example Ritchie, *Not the End of the World.*

15 See for example Murray Feshbach and Alfred Friendly, Jr., *Ecocide in the USSR: Health and Nature Under Siege* (New York: Basic Books, 1993).

16 Ashley Esarey, Mary Alice Haddad, Joanna I. Lewis, and Stevan Harrell, eds., *Greening East Asia: The Rise of the Eco-developmental State* (Seattle: University of Washington Press, 2020); Xianchun Tan and Henry Lee, "Comparative Assessment of China and U.S. Policies to Meet Climate Change Targets," Harvard Belfer Center for Science and International Affairs, accessed August 22, 2024, www.belfercenter.org; Yifei Li and Judith Shapiro, *China Goes Green: Coercive Environmentalism for a Troubled Planet* (Cambridge: Polity Press, 2020).

17 As one leading example, see China's moves to eliminate coal-based heating in many homes in the late 2010s, which showed little concern for the ability of poor households to afford alternatives to coal heat for their homes in the winter. For more general discussion, see Li and Shapiro, *China Goes Green.*

18 See for example, Garett Jones, *10% Less Democracy: Why You Should Trust Elites a Little More and the Masses a Little Less* (Stanford, CA: Stanford University Press, 2020).

19 Note that this paper was written in 2023. Recent events now make even more salient the general argument here that too much democracy enables deep harms to society, and that the best solution is to more securely insulate institutions that protect basic rights, regulate externalities, and protect the basic structure of well-regulated liberalism and capitalism from democratic and authoritarian distortion. It is an embarrassment that the past century of political philosophy has ignored the fundamental role of too much democracy in catastrophically undermining the universal freedom and prosperity that well-regulated liberalism and capitalism could otherwise provide.

20 See Battistoni, "Domination in the Age of the Externality," in this volume.

21 See for example Matthew Adler, *Measuring Social Welfare* (Oxford: Oxford University Press, 2019), and Matthew Adler and Marc Fleurbaey, *The Oxford Handbook of Well-Being and Public Policy* (Oxford: Oxford University Press, 2016). A central proposal within these studies for "equity weighting" within cost-benefit analysis of policy was endorsed by the updated guidance from within the US government in 2023; see "Circular A-4," November 9, 2023, The White House, accessed August 23, 2024, www.whitehouse.gov.

22 See for example Muller and Mendelsohn, *Using Marginal Damages in Environmental Policy.*

23 For some related discussion of the concept of well-regulated capitalism, and the importance of proper regulation to ethically desirable capitalist systems, see Budolfson, "Arguments for Well-Regulated Capitalism, and Implications for Global Ethics, Food, Environment, Climate Change, and Beyond."

24 For discussion of these differences, see for example Julia Nefsky, "Collective Harm and the Inefficacy Problem," *Philosophy Compass* 14, no. 4 (2019): e12587.

25 See for example, Sunstein, *The Cost-Benefit Revolution*, and Stavins, ed., *Economics of the Environment.*

5

THE CHICAGO SCHOOL'S COASEAN INCOHERENCE

MADISON CONDON

In her chapter in this volume, "Domination in the Age of the Externality," Alyssa Battistoni traces the concept of the "externality" across the past century: from welfare economist Arthur Pigou's "external economies" to Ronald Coase's "social cost."[1] She describes the Coasean project as one of removing the problem of domination from the harm of externalities by ignoring the underlying structural inequalities that shape who has the power to fight back against imposed harms. Coase does this by stripping the externality of Pigou's moral judgment: the assumption that pollution is bad. To Pigou, harms caused by pollution were left uncompensated, which necessitated non-market, government intervention. In "The Problem of Social Cost," Coase struck right at the heart of welfare economics: How was Pigou, or the government, to know how much people value pollution, or their health? Indeed, people observably choose to live near factories rather than move away.

What interests me about this cleansing transformation of the externality is that Coase had proven himself very capable of acknowledging the existence of domination in other aspects of the economy—indeed in the hierarchical nature of the large corporation. He observed that when an employee moves from one department to another, "he does not go because of a change in relative prices, but because he is ordered to do so."[2] Importantly, in Coase's original framing, these relationships of power were fundamentally different and *apart* from market relations, though influenced by them.[3]

My response to her chapter explores the ways Battistoni's argument interacts with Coase's *other* paper that had a major influence on the law and economics movement: "The Nature of the Firm." This article was published in 1937, twenty-three years before "The Problem of Social Cost." I ask what we can learn by connecting Battistoni's account of domination in externalities to Coase's earlier work on the corporation. By doing so, I hope to expand upon Battistoni's account of the role of both *intent* and *responsibility* in society. I argue, not uniquely, that corporations (or more precisely, the people that lead them and profit from them) have far more responsibility for the climate crisis than the rest of us. Battistoni discusses the many ways our choices are truncated as consumers, and our intentions are perverted. She argues it is for this reason that we must focus on systems rather than individuals as a force for change. I would like to focus our attention on the particular system comprised of multinational corporations and the *laws* that enable their destruction of our world.

Tracing the divergent academic interpretations of Coase and where their influence lands can reveal the law's inconsistent conception of just what a corporation is or should be. In antitrust law, the descendants of Coase highlighted the importance of firms for consumer welfare, as measured by keeping prices low. This interpretation enabled consolidation under the umbrella of a single firm while simultaneously prohibiting other forms of coordination, including through co-ops and unions.[4] But in corporate law, Chicago-style scholars took the position that Coase was wrong: The allocation of resources inside of a firm was ultimately directed by markets with or without the legal fiction of the firm boundaries.[5] This position was first useful in arguing against the corporate social responsibility campaigns of the late 1960s and early 1970s.[6] Eventually, the anti-Coasean firm helped solidify the legal norm of shareholder primacy.

I argue that these synchronous moves put all *moral responsibility* on the consumer, the constituent with the least amount of power, deflecting it from both the corporate manager and the institutional investor. Indeed, the law and economics movement helped to argue for limiting the choices and political power of shareholders over corporations, just as they insisted that profit maximizing was *for the shareholders*.[7] This move was to deny the power of the manager and the corporation. The ways in which the Chicago School interpreted this first Coase paper—and its clear acknowledgment of domination—can shed light on Battistoni's dilemma about moral choices mediated by markets. I argue for resurrecting pre-neoliberal legal conceptions of the corporation as a moral entity and a locus for political change in our fight against the climate crisis.

Nature of the Firm

Although it was published in 1937, Ronald Coase undertook most of the research and thought behind "The Nature of the Firm" during a 1931 tour of American corporations. As a twenty-year-old, he left England to visit the University of Chicago, which he used as a jumping off point to explore large midwestern manufactures.[8] He was interested in the relationship between General Motors and its suppliers. How did automobile companies decide whether to contract for a certain part or instead produce it in-house? In Coase's telling, part of the genesis for his "make or buy?" question was his early identification as a socialist—a remarkable footnote for someone who grew up to be one of the grandfathers of law and economics.[9] Lenin's plan to run Russia as "one big factory" had made an impression on Coase, as had the ongoing debate among European economists whether a centrally planned economy was even theoretically possible.[10]

Coase was not the only economist at the time interested in the ways a corporation was *not* a market. His article opens with the quoted observation that the economy was composed of "islands of *conscious power* in this ocean of unconscious cooperation."[11] Scholars were documenting just how much the corporation had grown and changed over the past century, and Coase's inquiry was an "attempt to accommodate economics to the rise of the vertically integrated firm."[12] Coase argued that using the market mechanism itself had costs, and that there were situations when bringing a certain production process within the limits of the firm would result in cost savings. This is where Coase's conception of "transaction costs" first appears. His prime example points to the efficiency gains of the employer-employee relationship over a contractual one.

To Coase, the leadership of the firm, termed the *entrepreneur*, played a special role in the economy. The managers' *act of directing resources* reaped efficiency gains over using the price mechanism. Coase defines "entrepreneur" as "the person or persons who, in a competitive system, *take the place of the price mechanism* in the direction of resources."[13] The firms' limits were defined by managerial skill, or "decreasing returns to management." At some point the firm became too big for the manager to efficiently control production through top-down command, and it became cheaper to outsource a function to the market. Although Coase wrote "The Nature of the Firm" contemporaneously with now-classic accounts of the relationship between corporations and shareholders, he ignored the existence of finance entirely.[14] In Coase's world the concept of *entrepreneur* seemingly merges the investor and corporate manager into a single entity.[15]

While initially ignored, "The Nature of the Firm" was rediscovered by the Chicago School in the late 1950s.[16] The Coasean corporation was used as a starting point to build new legal theories of antitrust that quickly grew to become today's mainstream neoliberal framing utilized by the courts. This reasoning led to the law looking more favorably on corporate consolidation. Academics argued that large firms led to productive efficiencies that redounded to the consumer, the most important constituent in the eyes of the new Chicago antitrust theory.[17] Simultaneously, however, a different group of law and economics scholars argued *against* the Coasean framing in corporate law.[18] They used an inverted version of Coase to argue that the firm was merely a "nexus of contracts," with nothing special about its hierarchical form or managerial structure.[19]

In the 1960s and '70s, the leadership and direction of the corporation became a center of public debate. A wave of consumer advocacy in the form of civil rights boycotts and Nader's Raiders launched one attack on the corporation as a political, and potentially harmful, entity in American society. The power of organized labor also experienced a brief resurgence.[20] It was at this time that another corporate constituency, the shareholder, began to emerge as an alternative force for social change.[21] Against these growing pressures, the "nexus of contracts" theory enabled the argument that the corporation was just as constrained by the market as everyone else—there was no managerial discretion to sway.

The idea that a firm was not an entity in its own right, but simply a "nexus of contracts," was taken up by Michael Jensen and William Meckling. They put forward a model of the firm in which all parties—workers, consumers, and shareholders alike—enter into voluntary contractual relationships with the legal fiction of the firm.[22] Under this theory, shareholders contracted to gain control over the firm in exchange for taking on the risk of being left holding the bag in the event of bankruptcy. This framing evaporates the hierarchical nature of the firm—the shareholder's residual profits and control power are just part of the contractual terms. This interpretation supported the idea that corporate directors owe their fiduciary duties to shareholders rather than some fictional idea of the corporation—and enabled the "corporate raiders" and layoffs of the 1980s. At the end of the decade, legal scholar William Bratton argued that as neoclassical theorists "purged conventionally conceived power relationships from the firm, they also pushed out the conventional morality that restrains the powerful."[23]

Everything, Everywhere, Externalities

Battistoni reaches a similar conclusion about the legacy of Coase's other famous paper, "The Problem of Social Cost." In language resembling Bratton's, Battistoni describes the Coasean project as one of removing the problem of domination from the harm of externalities. It does this by stripping the externality of Pigou's moral judgment: the assumption that pollution is bad. To Pigou, harms caused by pollution were left uncompensated, which necessitated non-market, government intervention. In "The Problem of Social Cost," Coase struck right at the heart of welfare economics: How was Pigou, or the government, to know how much people value pollution, or their health? Indeed, people observably chose to live near factories rather than move away.

Battistoni dwells on Coase's understanding of externalities as *mutual* costs. For Coase, the smokestack can only cause harm to health if people choose to live near it: "Both parties cause the damage."[24] Like many students and scholars, Battistoni finds this framing unsatisfying and digs deep into why: Those suffering from externalities can never come to the hypothetical bargaining table as equals against those benefiting from them. It is not simply a matter of the transaction costs of organizing, she argues, but of differences in power: Externalities are a form of domination.[25] In this way, those suffering environmental harms are an exploited class much like the labor class. The market acts to separate the receivers of benefits from encountering and understanding their attendant harms, relieving us of any moral duty to take responsibility for the far-away effects.[26]

Her description of the overwhelming feeling that comes with confronting the morality of consumer choice under supply chain capitalism is compelling and familiar.[27] Taking consumer responsibility "requires reevaluation of our seemingly trivial choices: what we eat; where we live; how we get around; what we wear." One could hope for an alternative world, in which one could trust that their shampoo ingredients were not produced through ecologically destructive means and exploitative labor practices. It is worth asking for a system on which these mental exercises to calculate far-off impacts are not required. As a consumer, as a shareholder. It's exhausting.[28]

While ideal, perfectly competitive, markets would theoretically work to create the tragic powerlessness Battistoni describes, that description does not fully match the world we currently inhabit. I push back on Battistoni here in pursuit of not letting people unfairly off the hook—and potentially identifying pressure points for change. In a Coasean firm at least, the internal non-market allocation of resources is fundamentally determined by power. What the CEO/investor wants to do mostly flies, unless the employees revolt.

Where the entrepreneur's authoritarianism is crucial in Coase's firm, the Chicago School embraced the Coasean conception of the firm as efficient, but flipped Coase on his head when it came to corporate law, insisting that markets governed both inside and outside the firm, the corporation being but a legal fiction, a "nexus of contracts." Sanjukta Paul makes the point that the Chicago School turns antitrust law, through Coase, into a concern only for consumers.[29] And it *makes the assumption for us*—as we all are consumers—that we would prefer the law to encourage low prices above all else. Under law and economics legal interpretations, the various policy motivations behind old statutes fell away. While the drafters may have intended the Sherman Act to consider the benefits of small businesses, worker power, and "local control," these were inappropriate and inefficient considerations according to judges influenced by the Chicago School.[30]

As my own scholarship has argued, the embrace of "shareholder primacy" and efficient markets ideology has resulted in these same *disempowering* assumptions being made in corporate law.[31] Jensen and Meckling's (anti-Coasean) idea of the corporation assumes that the only thing the shareholders want is an individual firm's stock price to go up. In general, the law now imputes that assumption into firm directors' legal duties.[32] While the law formally states that directors owe duties "to a corporation and its shareholders," courts regularly boil this duty down to share-price maximization. It was for this reason that Twitter's board of directors felt that they legally had to sell the company to Elon Musk, even if perhaps they thought it would be bad for the company—whatever that is.[33]

But every so often courts interpret directors' duties to be *to the corporation* rather than shareholders. In these cases, what is best for "the corporation" is *again* measured in terms of share price. In the context of climate change, I have argued that corporate law should allow large investors to pressure corporations into decarbonizing.[34] Employing a straightforwardly Coasean analysis myself, I argued that for a market portfolio—say, the S&P 500—the harm from future climate damages outweigh the benefits from fossil production over a long-term investment horizon. I put forward a crude cost-benefit analysis to argue that avoiding future lost profits would more than justify the present costs of foregoing fossil profits. A growing number of institutions and scholars have echoed my argument. Recently, the professional association of insurance actuaries in the UK criticized pension fund under-appreciation of climate damages. The group argued that because actuarial modeling expects devastating economic losses in the latter half of this century, decarbonization should become part of a fund manager's fiduciary duty, for "if we do not mitigate climate change, it will be exceptionally challenging to provide financial returns."[35]

Nevertheless, corporate scholars today insist that even if a majority of the diversified investors of a fossil-heavy firm *voted* to sacrifice profits in order to decarbonize, existing case law would bar firm leadership from doing what the shareholders want.[36] In this case, the court would interpret directors' duties to "the corporation" itself—*as measured* in terms of its share price and profits. This heads-I-win, tails-you-lose approach to interpreting directors' duties means that while formally, directors have a duty "to the corporation and its shareholders," in practice their duty is only to some abstract notion of a shareholder with a single, assumed, want.[37]

This is especially troubling when one pulls back the mask of institutional investors. Who are the shareholders of the big polluting corporations? It's often "us."[38] Officially, a significant amount of shareholder "control" is held in the endowments of our universities, the pension funds of public employees, and—increasingly since the 1980s—the 401(k)'s of the professional class.[39] Each month, employers all over America take a slice out of their workers' paycheck and hand it over to an asset manager, to be held in keeping for retirement. It is an intentionally constructed system in which American workers' pay is exchanged for ExxonMobil stock, for example, but shareholder *control* power is retained by a third party, like a Vanguard or a Fidelity. When viewed in the light of climate change and externalities, this structural separation of market returns from moral duty is significant. These asset managers "control" enough corporate shares to significantly change the direction of the economy, if they wanted to.[40] But of course, in the eyes of the law, asset managers are corporations too, without wants.

The market is something that is supposed to elicit our desires, but by looking into Coase's legacy we can see how market-based arguments have been used to simply limit our choices—simultaneously removing our responsibility and power. It becomes clearer just how much of trying to fight the market is an uphill battle, the price system is stacked against you—as a consumer, worker, and shareholder.[41] Law and economics enables the position that the corporation cannot carry any social responsibility, and that when "the moral and the efficient diverge," the fault lies "with the consumers who were unwilling to pay to realize their moral preferences."[42]

Low-Bar Common Good

In Pigou's analysis, smoke pollution from a factory is assumed to be bad, requiring state intervention to make the factory pay the full costs of pollution. Battistoni recounts how Coase reframed the externality as an economic exchange between two private parties, one that only justified government intervention in the circumstance where the costs of transacting were too high to overcome. Pigou's world required the government to make a normative decision: that the social negative of the smoke pollution was worse than the social benefits provided by the factory, including jobs. Coase's critique rested on the fact that the government had no basis on which to make this call: How was the government to know how much people valued their health? We know how people value things through observing their preferences in the market. With no market, according to Coase, you are left making paternalistic, normative calls about the relative value of good and bad effects.

It is true. It is hard, and arguably problematic, for a judge to impute desires to large groups of people. Where to draw the line about state correction of externalities is a question Pigou never resolves. But the Chicago School does exactly what it accuses progressive alternatives of doing—embracing a normative, "paternalistic," version of the common good. The market-based approach assumes we as shareholders only want the stock price to go up, that we as consumers only want the product price to go down, and that "domination is superior to coordination" as a way of organizing the economy.[43] All of this blossomed, somehow, from Coase's "The Nature of the Firm."

My favorite part of Battistoni's chapter is her clarity on the current system's burden of proof: "We do not all need to agree on a common good," she argues, "but we do need to be able to work toward the common purpose of maintaining a habitable planet."[44] One might think that not destroying the planet is a low bar. Yet it is a standard the "free-market" system is observably failing to meet. I would encourage the law to be open to the normative assumption that we would all like to get through this—investors and consumers alike.[45]

Revisiting Climate Change as an Externality

While Battistoni endeavors in good faith to confront Coase on his own terms, as a reader I could not help getting caught up in the many ways climate change is not properly characterized as an externality. The Inflation Reduction Act of 2022 is arguably the result of successful campaigns to dethrone the carbon tax as the "optimal" climate policy.[46] A new green industrial policy focuses on capital turnover and the replacement of fossil infrastructure with investment in new green investment. This Bidenomics approach, to draw from Ilmi Granoff, considers climate change to be "a substitution problem, not an externalities problem."[47] Indeed, mainstream economists now write that the problem of climate change should be thought of as both externalities and "innovation failures."[48]

This thinking can only go so far—there is currently no non-emitting substitution for a cow, you really do need to simply choose to not eat real meat. And the limits to this green tech investment solution may indeed have radical implications.[49] But, like economists, I am interested in these "innovation failures." While I retain skepticism about many climate hopes rooted in techno-optimism, I have been struck by how rapidly the window of possible technological solutions has shifted in recent years. The "hard to abate" sectors now seem eminently more abatable—construction timelines for green hydrogen pipelines are being scheduled. Serious questions about scalability remain, but *suddenly* carbon capture has emerged as not only a realistic but a necessary solution.[50] Why now? It just seems suspiciously in the nick of time to me, or a bit late.

I have a clear memory of a fully electric sedan visiting my elementary school in the 1990s. I am not the only one to wonder about what happened to it: a full-length documentary is devoted to the question of *Who Killed the Electric Car?* (2006).[51] Case studies analyze the US solar industry that could have been, had it not been destroyed by the wave of 1980s financialization. Max Jerneck argues that after large energy conglomerates bought up solar entrepreneurs, their interest in investing further in solar energy died as short-term financial profits took precedence over R&D and diversification.[52]

Pointing out these alternative futures is meant to give context to my resistance of Battistoni's emphasis on the unintentionality of externalities. She argues that climate change is not "the kind of effect we could *anticipate* until recently."[53] Here, I think acknowledging the role of the corporation is particularly useful. As someone who learned about climate change from a *Time* magazine article in the mid-1990s, I have long been fascinated by how much we *did* anticipate the effects of burning fossil fuels. A small group of people in particular, scientists and executives at ExxonMobil, anticipated the effects with startling precision in 1977.[54] The American Petroleum Institute and the fossil companies it represented had been warned by physicists in the 1950s.[55] Before that, as early as 1856 Eunice Foote postulated that burning fossils might warm the Earth, after her experiments revealed the heat-trapping nature of carbon dioxide.[56] That her discovery was lost until 2011 says something about power as well.[57]

The concept of an externality never fit perfectly with climate change and fossil fuels. When the United States set up a permit trading system in the 1990s to confront acid rain, the idea was to force companies to "internalize" the cost of sulfur dioxide. They had to pay money to emit it, which led to two main effects. One effect was that installing pollutant-removing scrubbers at the coal-burning plant made more economic sense. The second was that western coal was now cheaper than dirtier, east coast coal. The geographic location of "cleaner" coal deposits had been a key factor underlying the political economy of air pollution regulation for decades. The Acid Rain Trading program has largely been hailed as a success—and been repeatedly used as an example for the potential of carbon trading. But when it comes to carbon dioxide and fossil fuels . . . they are kind of the whole thing. We are burning ancient plants made of carbon. Carbon isn't appropriately thought of as some incidental, and unwanted, contaminant that you can filter out. This means that polluting less requires more than simply installing a filter or buying a more expensive type of fuel—it means *reinventing the entire energy system*. As Leah Stokes, Matto Mildenberger, and others have eloquently pointed out, as a political matter this means that industry is likely to be doubly entrenched: The losers are clearly demarcated, but who the winners will be is unclear.[58]

Elaborating on her concept of domination, Battistoni writes, "Capital is not only power over others' labor . . . *but power over investment and hence purposive action more broadly: over what is produced and how*. It is the power not only over the production of commodities, but over the production of environmental conditions, and increasingly, over the condition of the planet itself."[59] Under the Chicago School, "transaction costs" come to make up the entire friction of externalities—but within that category is the truth that the stakes are hopelessly tilted in favor of the polluter. In the post-atomic plastic age, corporations mastered the ability to bring *entire new compounds* into being—conjure chemicals never seen before, whose qualities and effects were unknown. How should the exposed citizenry know how much to pay to stop the production of the compound no one knows anything about?

Responsibility and Intent

The first Earth Day in 1970 was the largest collective demonstration in American history.[60] When founding coordinator Denis Hayes looks back on the environmental sentiment that had been building in the 1960s, he points to the feeling of living in the atomic age:

> When I was born, Strontium-90 didn't exist. By the time I was a teenager every living creature on the planet had Strontium-90 in its bones or its shells. That is a fairly profound change and we'd done it . . . This was the first generation that had acquired the power of a geophysical force that could force brand new radioactive substances to be disseminated throughout the entire planet.[61]

I played a film clip of his remark to my environmental law students this spring, as litigation unfolded against 3M and other corporations for the harms of "forever chemicals."[62] Scientists recently confirmed that "rainwater everywhere on Earth" is now unsafe to drink due to these pollutants.[63] These tiny man-made molecules are so slippery and pervasive they are found in the blood samples of most people.[64] They are also extremely harmful to human health, understudied, and continuing to be produced. While 3M executives may not have *initially* known about their products' harmful effects, the evidence is overwhelming that at some point they knew a lot about its toxicity, and worked to hide it.[65]

The briefly championed corporate governance ideas of the 1970s sought to instill moral concerns into economic direction. As Sarah Haan has unearthed, Milton Friedman's infamous *New York Times Magazine* castigation of the demand for "corporate responsibility" was responding to one movement in particular: "Campaign to Make General Motors Responsible."[66] A small group of shareholders was demanding that General Motors become more "accountable" to the public. They proposed that GM undergo a "social audit," disclosing information on environmental harm and racial discrimination, and appoint new board directors that could represent the views of environmental experts and community leaders. The board candidates proposed by "Campaign GM" were intentionally not white men.[67]

As Battistoni points out, that the allocation of "external" harms has been about power the whole time has been clear to those bearing the brunt.[68] Robert Bullard, "the father of environmental justice," described the siting of waste facilities near communities of color as "internal colonialism."[69] Bullard pointed out that it was not simply "market forces" that landed toxic waste sites near poor communities: People made intentional decisions to seek out those communities that would be unlikely to muster the political power to fight back.[70] Others have argued that even without evidence of intentionality, the racial power dynamics enabling the "market" harm remain.[71] Indeed, environmental justice scholars have been making this point about markets hiding morals for decades. Tseming Yang, for example, has argued that a key failing of environmental law's economic utilitarianism is its inability to see minority oppression and the exercise of power.[72]

Corporations *shape* markets, and can create them entirely. 3M was only able to bring massive amounts of forever chemicals into the world because it worked to create a corporate demand for it. The power to decide what gets invested in, and what does not, holds power to mold the entire future, and constrain future alternatives. The objects that get produced are unfairly described as emanating from collective consumption demands.[73] As young Coase perhaps only semi-understood, corporations are a locus of *planning*.[74] But only a small number of people actually make the decisions—it is *their* version of the future that gets investment. Indeed, Iris Marion Young herself describes the acceptance of a profit-maximizing corporation as the result of a New Deal bargain between capitalists and workers.[75] Workers were promised rising wages and the welfare state in exchange for relinquishing debate about "control over production processes, or investment priorities."[76] In this bargain, distributive issues are the sole focus, while questions about alternative decision-making processes disappear. The corporation becomes depoliticized.

Conclusion: External to What?

Returning again to the origins of the externality, we can see the late 1960s and early 1970s as a pivotal era. People were waking up to the collective costs of industrialization and pushing back against corporate power. Against this democratic wave, the writings of the Chicago School worked to separate one human person into her different roles in the economy—consumer, worker, shareholder. They used the law to solidify the *divergent* interests of these roles, even as they preached the gospel of shareholder democracy and personal choice.

If the chemical progeny of 3M is in all of our drinking water and all of our blood, it is also in the water and blood of the people who control 3M. Like climate change, forever chemicals are challenging to categorize as a simple externality, implying that their harms could ever have been properly "internalized" through a price mechanism. As Battistoni predicts may happen if lawsuits successfully press fossil companies to pay for their climate harms, 3M the company itself may not survive the litigation. Bruno Latour diagnoses our current political moment as resulting from the collective decision of the elite not "to share the earth with the rest of the world."[77] He argues that we cannot overcome climate change until we learn to "live in the same world, . . . face up to the same stakes."[78] To Latour, the elite are only able to maintain this position through self-delusion. They think they can "escape from the problems of this planet by moving to Mars, or teleporting themselves into computers."[79]

And this is the final way that thinking about climate change through the market and through externalities fails to be useful. It may make perfect sense, but it does not persuade. The last time America had a mass ecological awakening, an unfortunate amount of political energy was focused on population growth as the root of environmental destruction. Garrett Hardin's "Tragedy of the Commons" is predominantly an argument for forced population control.[80] This sentiment of *not wanting to share* distracted from other ways of seeing the problem. Fifty years later, the 0.01% wealthiest humans are responsible for more emissions than the poorest 50%.[81] As the science mounts about the *collective* nature of our existential stakes, at some point it is not worth convincing the powerful that they are on Spaceship Earth with us. At some point you have to grab the controls.

Acknowledgments

I am grateful for the opportunity to comment on Alyssa Battistoni's wonderful chapter. The piece exposed me to new writers and opened new ways to read familiar ones. This chapter benefited from feedback from Luke Herrine and Eric Orts as well as from presentation at Northeastern University School of Law's Faculty Workshop and the American Society for Political and Legal Philosophy's conference on Climate Change.

Notes

1 Alyssa Battistoni, "Rethinking Domination in the Age of the Externality," this volume; A. C. Pigou, *The Economics of Welfare* (London: Macmillan, 1920); R. H. Coase, "The Problem of Social Cost," *Journal of Law & Economics* 3 (October 1960): 1–44.

2 R. H. Coase, "The Nature of the Firm," *Economica* 4 (November 1937): 386–405, at 387.

3 Sanjukta Paul, "Antitrust as Allocator of Coordination Rights," *UCLA Law Review* 67 (2020): 378–431, at 420 ("Managerial hierarchies, and the separation of work from ownership, were thus basic to Coase's account.")

4 Paul, "Antitrust as Allocator of Coordination Rights."

5 Michael Jensen and William Meckling, "Theory of the Firm: Managerial Behavior, Agency Costs, and Ownership Structure," *Journal of Financial Economics* 3, no. 4 (1976): 305–360, at 311 ("Viewed this way, it makes little or no sense to try to distinguish those things that are 'inside' the firm [or any other organization] from those things that are 'outside' of it.").

6 Milton Friedman, "The Social Responsibility of Business is to Increase Its Profits," *New York Times*, September 13, 1970.

7 Sarah C. Haan, "Civil Rights and Shareholder Activism: SEC v. Medical Committee for Human Rights," *Washington & Lee Law Review* 76, no. 3 (2019): 1167–1229, citing, e.g., Henry G. Manne, "The Myth of Corporate Responsibility or Will the Real Ralph Nader Please Stand Up," *Business Lawyer* 26, no. 2 (1970): 533–539, at 539.

8 Often by just showing up and asking to chat. R. H. Coase, "Nature of the Firm: Origin," *Journal of Law, Economics, & Organization* (1988): 3–17; see also Leigh Phillips and Michal Rozworksi, *Peoples Republic of Walmart* (New York: Verso, 2019), 49.

9 Coase, "Nature of the Firm: Origin"; Paul, "Antitrust as Allocator of Coordination Rights" (using "make or buy" terminology).

10 Coase, "Nature of the Firm: Origin"; Coase, "Nature of the Firm," citing E.F.M. Durbin, "Economic Calculus in a Planned Economy," *Economic Journal* 46, no. 184 (1936): 676–690.

11 Coase, "Nature of the Firm," citing D. H. Robertson, *Control of Industry* (1932). The quote famously continues ". . . like lumps of butter coagulating in a pail of buttermilk."

12 Barry Lynn, *End of the Line: The Rise and Coming Fall of the Global Corporation* (New York: Doubleday, 2005), 136.

13 Coase, "Nature of the Firm," 388n2. ("We can best approach the question of what constitutes a firm in practice by considering the legal relationship normally called that of 'master and servant' or 'employer and employee.'") Italics added.

14 Cf. Adolf Berle and Gardiner Means, *The Modern Corporation and Private Property* (New York: Macmillan, 1933).

15 In contrast, Gardiner Means recognized that as the corporation grew in size, so did its shareholder base. He worried that this dispersed shareholder class could not supervise these managers of corporations entrusted with their investment. Means argued that not only the "separation of labor from control" identified by Coase, but also the separation of ownership from control enabled the expansion of firm size. The corporate managers gained dominion over larger areas of economic control just as they lost accountability to investors. Warren Samuels and Steven G. Medema, "Gardiner Means's Institutional and Post-Keynesian Economics," *Review of Political Economy* 1, no. 2 (1989): 163–191, citing G. C. Means, "The Corporate Revolution," manuscript of submitted doctoral dissertation, Harvard University, Gardiner C. Means Papers, Series I, Franklin D. Roosevelt Library.

16 George L. Priest, "Ronald Coase, Firms, and Markets," *Man and the Economy* 1, no. 2 (2014): 143–157.

17 Paul, "Antitrust as Allocator of Coordination Rights."

18 Armen A. Alchian and Harold Demsetz, "Production, Information Costs, and Economic Organization," *American Economic Review* 62, no. 5 (1972): 777–795.

19 Jensen and Meckling, "Theory of the Firm: Managerial Behavior, Agency Costs, and Ownership Structure," 311 ("The private corporation or firm is simply one form of *legal fiction which serves as a nexus for contracting relationships.*").

20 Jefferson Cowie, *Stayin' Alive: The 1970s and the Last Days of the Working Class* (New York: New Press, 2010).

21 William Bratton, "The Separation of Corporate Law and Social Welfare," *Washington & Lee Law Review* 74, no. 2 (2017): 767–790.

22 Jensen and Meckling, "Theory of the Firm: Managerial Behavior, Agency Costs, and Ownership Structure."

23 William Bratton, "The 'Nexus of Contracts' Theory of the Corporation: A Critical Appraisal," *Cornell Law Review* 74 (1989): 407–465, at 455.

24 Coase, "The Problem of Social Cost."

25 Battistoni's critique draws from institutional economist K. W. Kapp's argument "that the struggle over the burden of social costs is better characterized in terms of struggle between classes with disparate power than as a market exchange between equal individuals." Battistoni, "Domination in the Age of the Externality"; see Karl William Kapp, *The Social Costs of Private Enterprise* (New York: Schocken Books, 1950). She argues that Kapp and Coase are aligned in recognizing the *reciprocity* of social costs. I note that Kapp's thinking here resembles that of legal realist Robert Hale. Hale argued that the market itself was a system of "mutual coercion": There was no such thing as the "free" market when it relied on an underlying state police power to enforce property rights. See, e.g., Robert Hale, "Coercion and Distribution in a Supposedly Non-Coercive State," *Political Science Quarterly* 38, no. 3 (1923): 470–494. While Hale is not well known, a new group of legal scholars has resurrected his lessons for combatting the legacy of the Chicago School. See, e.g., Sabeel Rahman, "Law, Political Economy, and the Legal Realist Tradition Revisited," *LPE Blog*, October 29, 2018, www.lpeproject.org. The last time his work was of interest to legal scholars was the pivotal turning point of the 1960s into the '70s, when he was lauded as "an originator of the concept of 'private government.'" Arthur S. Miller, "Private Governments and the Constitution," in *The Corporation Take-Over*, ed. Andrew Hacker (New York: Harper & Row, 1964): 122–149, at 138–139; see also Warren J. Samuels, "The Legal Economics of Robert Lee Hale," *Miami Law Review* 27, nos. 3–4 (1973): 261–371.

26 Battistoni, "Rethinking Domination in the Age of the Externality" ("that markets *constitutively* detach intentions from consequences, at both the individual and collective level."). Italics in original.

27 See also Anna Tsing, "Supply Chains and the Human Condition," *Rethinking Marxism* 21, no. 2 (2009): 148–176, at 158, italics in original ("Supply chains *depend* on those very factors banished from the economic; this is what makes them profitable.")

28 I am reminded of the television show *The Good Place*, wherein a point system determines whether someone gets into heaven, tabulating the good and bad actions taken over their lifetime. In modern times, no one earns enough points to escape hell: "Life is now so complicated, it's

impossible for anyone to be good enough for the Good Place. These days, just buying a tomato at a grocery store means that you are unwittingly supporting toxic pesticides, exploiting labor, contributing to global warming. Humans think that they're making one choice, but they're actually making dozens of choices they don't even know they're making!" Kate Yoder, "'The Good Place': It's Hard to Be Good When the World's on Fire," *Grist*, January 26, 2020, www.grist.org.

29 Paul, "Antitrust as Allocator of Coordination Rights."

30 Paul, "Antitrust as Allocator of Coordination Rights." See also Lina M. Khan, "Amazon's Antitrust Paradox," *Yale Law Journal* 126, no. 3 (2017): 710–805.

31 Madison Condon, "Externalities and the Common Owner," *Washington Law Review* 95, (2020): 1–81; Madison Condon, "Climate Change's New Ally: Big Finance," *Boston Review*, July 28, 2020.

32 See generally, William Allen, "Our Schizophrenic Conception of the Business Organization," *Cardozo Law Review* 14, no. 2 (1992): 261–281.

33 Ann Lipton, "Every Billionaire Is a Policy Failure," *Virginia Law & Business Review* 18, no. 3 (2024): 327–433.

34 Condon, "Externalities and the Common Owner."

35 Sandy Trust, Sanjay Joshi, Tim Lenton, and Jack Oliver, "The Emperor's New Climate Scenarios," *Institute and Faculty of Actuaries*, July 2023, www.actuaries.org.uk.

36 Marcel Kahan and Edward Rock, "Systemic Stewardship with Tradeoffs," *Journal of Corporation Law* 48, no. 3 (2023): 497–539; cf. Jeffrey N. Gordon, "Systematic Stewardship: It's Up to the Shareholders: A Response to Profs. Kahan and Rock," *Journal of Corporation Law* 48 (2023): 26–31.

37 See also Lipton, "Every Billionaire Is a Policy Failure," 409 (equating the upholding of share price primacy as "vindicating the rights of an abstract notion of shareholder").

38 Jim Kane, "Here for a Good Time, Not a Long Time: Asset Managers at the Infrastructure Party," *New Labor Forum* 32, no. 3 (2023): 104–107, at 107 ("It is not a stretch to say we are negotiating against ourselves and it should not be a stretch to imagine returning to a system that cuts that circuit.").

39 Benjamin Braun, "Fueling Financialization: The Economic Consequences of Funded Pensions," *New Labor Forum* 31, no. 1 (2022): 70–79.

40 Condon, "Externalities and the Common Owner."

41 Cf. Battistoni, "Rethinking Domination in the Age of the Externality," pointing out that she cannot pay more for something that causes environmental harm even if she values the environment more than the market does.

42 Robert T. Miller, "The Coasean Dissolution of Corporate Social Responsibility," *Chapman Law Review* 17, no. 2 (2014): 381–411, at 410.

43 Paul, "Antitrust as Allocator of Coordination Rights"; see also Jedediah Britton-Purdy, David Singh Grewal, Amy Kapczynski, and K. Sabeel Rahman, "Building a Law-and-Political-Economy Framework: Beyond the Twentieth-Century Synthesis," *Yale Law Journal* 129, no. 6 (2020): 1784–1835.

44 Battistoni, "Rethinking Domination in the Age of the Externality."

45 Cf. Trust et al., "The Emperor's New Climate Scenarios"; Amelia Miazad, "From Zero-Sum to Net-Zero Antitrust," *UC Davis Law Review* 56 (2023): 2067–2103.

46 Leah C. Stokes and Matto Mildenberger, "The Trouble with Carbon Pricing," *Boston Review*, September 24, 2020.

47 Ilmi Granoff (@theilmatic), ". . . the core economic problem of decarbonization is not negative externality, but technological substitution." Twitter (now X), April 12, 2023, twitter.com/theilmatic/status/1646281472687640576. On "Bidenomics," see Harold Meyerson, "It's Hamiltonian! It's Jeffersonian! It's Bidenomics!" *American Prospect*, July 3, 2023.

48 Sarah C. Armitage, Noël Bakhtian, and Adam B. Jaffe, "Innovation Market Failures and the Design of New Climate Policy Instruments," NBER Working Paper No. 31622, August 2023.

49 But I am going to unsatisfyingly table the question for the purposes of this chapter.

50 I am mostly attempting to summarize received wisdom in the mainstream press. But I also write as someone who was required to take a class devoted substantially to carbon capture technology in the mid-2000s and have watched the industry over the decades. I have been very surprised by how much costs have dropped on technologies such as Direct Air Capture.

51 *Who Killed the Electric Car?*, directed by Chris Paine (Electric Entertainment, 2006).

52 Max Jerneck, "Financialization Impedes Climate Change Mitigation: Evidence from the Early American Solar Industry," *Science Advances* 3, no. 3 (2017): e1601861.

53 Battistoni, "Rethinking Domination in the Age of the Externality."

54 G. Supran, S. Rahmstorf, and N. Oreskes, "Assessing ExxonMobil's Global Warming Projections," *Science* 379, no. 6628 (2023): eabk0063.

55 Benjamin Franta, "Early Oil Industry Knowledge of CO2 and Global Warming," *Nature Climate Change* 8 (2018): 1024–1025.

56 Eunice Foote, "Circumstances Affecting the Heat of the Sun's Rays," *American Journal of Science and Arts* 22, no. 66 (1856): 382–383.

57 Katherine Wilkinson, "Why We Can't Forget Eunice Foote's Science Work," *Time*, July 17, 2019.

58 Stokes and Mildenberger, "The Trouble with Carbon Pricing."

59 Battistoni, "Rethinking Domination in the Age of the *Externality*." Italics added.

60 It was arguably surpassed by the George Floyd and Black Lives Matter protests, ignited in 2020 and ongoing. Larry Buchanan, Quoctrung Bui, and Jugal K. Patel, "Black Lives Matter May Be the Largest Movement in U.S. History," *New York Times*, July 3, 2020. The 1970 Earth Day remains the largest single-day demonstration across multiple cities.

61 *Earth Days*, directed by Robert Stone (PBS: 2010), www.pbs.org.

62 Lisa Friedman and Vivian Giang, "3M Reaches $10.3 Billion Settlement in 'Forever Chemicals' Suit," *New York Times*, June 22, 2023.

63 Rosie Frost, "Rainwater Everywhere on Earth Unsafe to Drink Due to 'Forever Chemicals,' Study Finds," *EuroNews*, August 4, 2022, www.euronews.com.

64 Tom Perkins, "PFAS Left Dangerous Blood Compounds in Nearly All US Study Participants," *Guardian*, September 29, 2022.

65 Investigative reporter Sharon Lerner has an extensive body of work on this. See, e.g., Sharon Lerner, "3M Knew About the Dangers PFOA and PFOS Decades Ago, Documents Show," *The Intercept*, July 31, 2018, www.theintercept.com.

66 Sarah Haan, "Exclusion in Corporate Law," in *Oxford Handbook of Corporate Law and Governance*, 2nd edition, ed. Jeffrey N. Gordon and Wolf-Georg Ringe (Oxford University Press, forthcoming).

67 Donald E. Schwartz, "Proxy Power and Social Goals—How Campaign GM Succeeded," *St. John's Law Review* 45, no. 4 (1971): 764–771 (". . . it is not sufficient that the Board include persons who have concerned themselves with problems of black people; the Board needed a black person. The same is true of a woman, as we develop an increased consciousness of discrimination against women.").

68 Battistoni, "Rethinking Domination in the Age of the Externality."

69 Robert D. Bullard, "Anatomy of Environmental Racism and the Environmental Justice Movement," in *Confronting Environmental Racism: Voices from the Grassroots*, ed. Robert D. Bullard (Boston: South End Press, 1993): 15–39.

70 Bullard, "Anatomy of Environmental Racism."

71 Sheila Foster, "Justice from the Ground Up: Distributive Inequities, Grassroots Resistance, and the Transformative Politics of the Environmental Justice Movement," *California Law Review* 86 (1998): 775–841.

72 Tseming Yang, "Melding Civil Rights and Environmentalism: Finding Environmental Justice's Place in Environmental Regulation," *Har-

vard *Environmental Law Review* 26, (2002): 1–32. ("In essence, the market dynamics perspective makes it that much easier for environmentalists, government regulators, and industry to absolve themselves from responsibility for addressing societal discrimination and social inequities in addressing the causes of environmental problems.")

73 See also Luke Herrine, "What Is Consumer Protection For?," *Loyola Consumer Law Review* 34, no. 2 (2022): 241, 274 (arguing against the flawed concept of "consumer sovereignty" by recognizing that consumer markets are not neutral, but rather are "made up of a series of collectively constructed institutions that shape which options are available").

74 Cf. Jamie Medwell, "The Lucas Plan Was a Workers' Alternative to Neoliberalism," *Tribune* July 3, 2022, tribunemag.co.uk; Rob Walker, "Eco-Pioneers in the 1970s: How Workers Tried to Save Their Jobs—and the Planet," *Guardian*, October 14, 2018, www.theguardian.com.

75 Battistoni's essay is guided by Iris Marion Young's work on structural domination and the "institutional conditions" that shape our actions. Battistoni, "Rethinking Domination in the Age of the Externality," citing, e.g., Iris Marion Young, *Responsibility for Justice* (Oxford: Oxford University Press, 2011), 11.

76 Iris Marion Young, *Justice and the Politics of Difference* (Princeton, NJ: Princeton University Press, 1990), 71.

77 Bruno Latour, *Down to Earth*, trans. Catherine Porter (Cambridge: Polity Press, 2018).

78 Latour, *Down to Earth*, 25.

79 Latour, *Down to Earth*, 30.

80 Andrew Follett, Brigham Daniels, and Taylor Petersen, "The Tragedy of Garrett Hardin's Commons," in *Cambridge Handbook of Commons Research Innovations*, ed. Sheila Foster and Chrystie Swiney (Cambridge: Cambridge University Press, 2021): 26–33 ("Hardin's main point (and the subject of the bulk of his article) is that, because the world is becoming overcrowded, controlling the human population requires a reexamination of individual liberties. This is not an interpretive reach—it is his stated thesis.").

81 Jared Starr, Craig Nicolson, Michael Ash, Ezra M. Markowitz, and Daniel Moran, "Income-Based U.S. Household Carbon Footprints (1990–2019) Offer New Insights on Emissions Inequality and Climate Finance," *PLOS Climate* 2, no. 8 (2023): e0000190.

PART III

CLIMATE CHANGE AND MORAL RESPONSIBILITY

6

ON THE MORAL CHALLENGE OF THE CLIMATE CRISIS

LUCAS STANCZYK

When the planes were grounded by the virus, the fires in Siberia were only getting started. Temperatures in the Arctic would soon break a hundred degrees Fahrenheit. Earth frozen since the beginning of history began smoldering over a colossal area. Within weeks, millions of acres of forest and tundra had been burned up by the flames. Even so, the fires were but a single link in a chain of unparalleled events. The following year, warm water was found beneath the massive Thwaites glacier in Antarctica. Like the active methane leaks that have now been documented on the seabed, this discovery is bound to alter when we will be told to expect even more rapid melting. We know this because it has happened so regularly and so often. With every decade, the poles have been warming faster than the outlying predictions of the previous period. Where once we thought it would take a century, we are now told that the Arctic may be ice-free as early as the 2040s. Less ice will lead to yet more ice-destroying rain, evaporation produced by the sea absorbing more of the sun's rays. In turn, more wind will stir the seas and the methane deposits trapped beneath, while rains that reach the land will accelerate the thawing of the frozen ground. As the northern climate warms, the annual fires will reach still more of the boreal forests, releasing enormous quantities of carbon stored in vegetation, peat, and humus. Predictably, there will be more melting, more rain and flooding, more drought and fire, and still more warming. All the while, the climate will not care if the people are distracted by the growing jingoism or the latent fascism or the failure to respond rationally to a virus. On the contrary, with every hurricane, drought, and crop failure, nature will present humanity with yet more occasions for division and further obstacles to solidarity.

Until recently, most thinking people were as good as sleeping on this issue. To be sure, most had heard of its existence at some point, and many even saw through the denialist propaganda of their bought-and-paid-for politicians. Yet for decades, even privileged people in wealthy countries persistently ranked the climate crisis near the bottom of their so-called policy priorities. It is a ray of hope in a time of darkness that majorities of voters are slowly beginning to revise what they profess to be their sovereign preference on this matter. Nevertheless, I believe that even well-meaning people continue to be deluded by the scale of the challenge that we face. In a word, it's not the case that all we need is the political will to pass a Green New Deal. Aside from distorting the politics that produced the climate crisis in the first place, this familiar invocation ignores both the technological challenge and the moral difficulty of conceiving a suitably rapid energy transition. Overcoming these blind spots is essential to understanding the strategic predicament of everyone who wishes to avert the suffering and the cruelty now looming in our future.

When I first started thinking about this subject, I thought the philosophical dimensions were interesting but quite irrelevant practically speaking. Like others, I was content to know that the world just needed to cut its greenhouse gas emissions. As I learned more about the problem, I gradually lost my unexamined confidence. In fact, I went from thinking that we already know what we morally ought to do about the climate crisis to realizing that we don't even have an adequate approach to this question, let alone a fully defensible answer. One reason why it's hard to see this is that the consequences of past inaction are closer and more dire than many people think, even while the challenge of avoiding worse consequences is now greater than most of us suppose. There is certainly a large number of low-hanging fruit still left to be picked, stoves to be replaced, and houses insulated. But the changes that will make most of us better off, if only the politics allow, are not where any of the difficult moral problems lie.

The problems surface only once we're honest about the remaining carbon budget and the difficulty of a global energy transition.

Even ignoring for a moment the prison of our capitalist politics, the world is very far away from being able to decarbonize all existing human energy consumption. Hence, total consumption would need to drop for humanity to stop adding greenhouse gases to the atmosphere in the near term. Meanwhile the clock is ticking on feedback mechanisms and other catastrophic outcomes. This means that we cannot think responsibly about the climate crisis without evaluating both per capita energy use and global population size. Yet these subjects are immensely difficult and certainly make a mockery of the conventional policy approaches. They also raise hard questions concerning how much we owe to future generations and what we owe to other animals.

In this chapter, I will present an approach that I think is the only satisfactory way of dealing jointly with the first three problems: (i) evaluating restrictions on carbon-intensive present-day consumption (ii) as requirements of intergenerational justice (iii) without succumbing to the problems of population ethics. On this approach, duties of intergenerational justice are owed first and foremost by the adult powerholders in the present generation to the younger people who are already alive and will replace them someday. The core imperative of intergenerational justice on this view is to put in place all those restrictions on existing rights and freedoms that, unless they are put in place, will require even more serious sacrifices having to be made later by today's younger people, lest they fall afoul of the very same imperative in the future. In turn, the essence of intergenerational *in*justice is a form of "kicking the can down the road": *failing* to put in place restrictions on existing rights and freedoms that are necessary to prevent even more serious sacrifices having to be made later for the purpose of enacting justice and securing every living person's rights to be free from avoidable famine, drought, resource wars, and so on.

I shall argue that this is exactly what today's most powerful people are unconscionably doing now, and it is exactly how you and I were wronged by the people who were in charge 30, 40, 50 years ago. They failed to make changes to energy systems that, though they would have been costly and socially disruptive, would have been much less burdensome to undertake in the time that was then still available to prevent catastrophic climate change. Instead, they chose to indulge much less urgent interests in consumption and accumulation and left us in a much more dire situation. In turn, today's powerholders are doing the very same thing to the children born around the world yesterday. They are indulging various less important interests in accumulation and consumption rather than reducing emissions and accelerating the global energy transition, ensuring that even more burdensome steps will later be morally required—of these children, for this purpose—else the world eventually will collapse from catastrophic climate change.

I will suggest (because I have come to think) that this approach is the only satisfactory way of reasoning about the intergenerational dimension of climate change. Even so, setting out a way of reasoning does not itself answer urgent practical questions. Moreover, there are dimensions of the climate crisis that are absolutely central to the question of what our governments presently ought to do but regarding which, after many attempts to think them through, I still hardly know where to begin. In the final part of this chapter, I will bare my ignorance about two such issues before focusing in more detail on a third. The first issue is the question of what it would be for our generation to address the climate crisis in a way that respects the claims, not merely of human beings, but of all the other wild and domesticated animals. The second question is how to make fateful social decisions regarding the pace of the global energy transition under conditions of profound uncertainty about the consequences of these actions. The final issue is how you and I ought to reason about our moral obligations when we admit to ourselves that the powerful will, predictably, not do any of these things, and that other people will respond to the fallout with still more wrongdoing in their turn.

The seas will warm, the storms will come, the insurance markets will fail; and yet the people who fly private jets will more or less be made whole. And then one year, the staple crops will wither simultaneously in many highly populated places around the world. That's when the reactionary impulse against the growing migrant flows will really start to take hold. Think about what just a few million foreigners on the shores of Europe and the US southern border have done to the politics of these countries. Children in cages; openly fascist parties; EU-funded goons beating unarmed migrants. If the recent past is any indication, the response to hundreds of millions of Asian, African, and Latin American migrants will be even more toxic. How should such sober expectations affect how you and I, here and now, understand our intellectual and political obligations? Needless to say, this final dimension of the climate crisis—the non-ideal theory of climate ethics—contains all of the most urgent practical questions. What's more, these questions do not in any way avoid the challenges of so-called ideal theory; on the contrary, they inherit those philosophical problems while adding new ones. Thanks to these compounding difficulties, I no longer think that it's already perfectly clear what we, together and individually, should do. Instead, I have come to think that climate change is perhaps the hardest moral problem that humanity has ever faced, to the extent that this phrase (a problem that "humanity has faced") even makes sense in a still-unplanned, uncoordinated, and divided world.

Understanding Our Predicament

To appreciate the difficulty, it helps to begin by reflecting on a familiar kind of hard-nosed skepticism. On this view, thinking about the ethical dimensions of the climate crisis is at best irrelevant and at worst a dangerous distraction. The putative reason is that states don't care about ethical considerations when they set their policies; they care only about advancing their self-interest. This general proposition is said to be borne out by the shape of the existing energy transition. Europe, the continent that has made the most progress toward decarbonization, is also the continent that is least rich in oil and gas deposits. The United States began to leave coal behind only when fracking technology unleashed its awesome natural gas reserves; not coincidentally, it has since become the world's single largest producer of both natural gas and oil. China is rapidly developing its western coal reserves so that it can reduce its dependence on oil from overseas, not least because such dependence makes the regime vulnerable to a future naval blockade over the issue of Taiwan. Russia, as the third largest oil producer, is entirely unconcerned about climate change and is planning drilling projects in the Arctic even as it melts. Canadian politicians talk politely about the future of our children while developing the second largest and by far the dirtiest oil deposit in the world. And most other fossil fuel–producing countries are similarly disguised climate wolves. So, the thought goes, it is better not to talk about ethics if one wants to see meaningful global greenhouse gas reductions. Instead, one should speak to the self-interest of states and ask three simple questions. Who stands to benefit most from a robust climate treaty? Who stands to benefit less? And how can the less vulnerable countries be bought off?

Taking their own advice, the economists Eric Posner and David Weisbach report that "an optimal climate treaty could well require side-payments to rich countries like the United States and rising countries like China, and indeed possibly from very poor countries which are extremely vulnerable to climate change—such as Bangladesh."[1] They call this the "realist" approach to international negotiations. I begin with this approach because, aside from being quietly influential in policy discussions, it is also instructively naïve about the nature of the climate crisis. Most importantly, the proposal assumes that it is in the narrow economic self-interest of the vulnerable countries to take rapid action to stave off the worst effects of climate change. But that is a mistake, based on a failure to appreciate two things: the very temporally extended consequences of the growing planetary energy imbalance, and just how difficult it will be to avoid irreversible tipping points in the near term whose worst effects will come only in the distant future.

There are many such tipping points in the climate system, and each has the potential for devastating consequences.[2] They include accelerated melting of the Greenland and Antarctic ice sheets, accelerated melting of the polar sea ice, accelerated warming from the thawing of the permafrost, accelerated warming from the growing fires in the boreal forests, the collapse of the Amazon rainforest and the permanent loss of a planetary carbon sink, warmer southern oceans from the collapse of the Atlantic overturning circulation system, the collapse of marine food webs upon the death of the remaining coral reefs, the tipping of the Pacific Ocean into a permanent El Niño state, hotter conditions in the Northern Hemisphere leading to yet more fires in the boreal forests, and yet more melting of the polar ice and more methane release from soil decomposition. Moreover, some of these climate system shifts may well be breached in the near future. Most alarmingly, the Amazon may be near the point of an irreversible collapse into savannah,[3] and the Atlantic overturning circulation (which drives many of the world's weather-affecting ocean currents) could pass the point of no return as early as 2025.⁴ This last shift alone may devastate agriculture in Central America and parts of Europe while catastrophically shifting the African monsoon.⁵

However, none of these things mean that it is in the narrow self-interest of the vulnerable countries to take the steps that would be necessary to stave off the worst effects of climate change. That's because the worst effects of climate change—all of the world's coastal cities being swallowed by the oceans, the massive expansion of deserts outward from the equatorial latitudes, heat waves everywhere that are not survivable by mammals—will take many centuries to unfold, long after everyone alive today is dead. Meanwhile, the near-term challenge of avoiding temperatures at which tipping elements may be breached is hard to overstate. Consider for instance what it would take to stop the world from warming more than 1.5 or 2 degrees.⁶

As of January 2023, the carbon budget for a 50% chance of not exceeding 1.5 degrees was only 250 gigatons of CO_2, or just over six years of current global emissions. For a somewhat higher 66% chance of limiting warming to 1.5 degrees, the carbon budget fell to less than four years of current annual emissions. This means that, at this point, the only way to avoid dangerously higher temperatures is for annual emissions to go into a sudden freefall. What would be required for this purpose would be instantly ceasing the use of virtually all fossil fuels: all marine diesel used for transoceanic shipping; all existing jet fuel; all of the gasoline and heating oil; all of the coal that drives the world's steel plants and cement kilns; all of the natural gas that is used to make the world's fertilizer. Even then, the global temperature would still rise by more than 1.5 degrees above the preindustrial level as the sulfur pollution from existing coal plants fell to the ground and removed a major atmospheric albedo cooling effect.⁷ The recent warming of the Atlantic Ocean stemming partly from reductions in marine diesel sulfur content is a prelude to the spike in global temperatures that the world can expect to see on the loss of the global sulfur albedo.⁸

Now consider the challenge of avoiding more than 2 degrees of warming. As of 2023, at most 950 gigatons of CO_2 could be emitted to leave the planet with a 66% chance of avoiding more than 2 degrees of warming by the year 2100. This carbon budget amounts to less than twenty-four years of current annual emissions. However, if the world's emissions continue at the present rate *for just another ten years*, then the slope of emissions reductions required to stay within 2 degrees of warming will be *just as steep as* the freefall required for a chance of staying under 1.5 degrees today. While this observation may be striking, it is just a consequence of the most important truth of global warming. The global temperature will keep rising—to an ever-higher plateau above 2 degrees—until humanity wholly transforms its land-use practices and stops the use of fossil fuels entirely. Merely "reducing" planet-warming emissions is not going to be enough. The global temperature will continue rising as long as humanity is adding anything whatsoever to the atmospheric stock of greenhouse gases. And yet the world's emissions are still growing rather than holding steady, let alone falling. The uptrend was briefly interrupted when the COVID-19 pandemic shut down the global economy in 2020. But then emissions rebounded strongly in 2021 and hit another all-time high the following year. In short, the prospect of avoiding temperatures that could trigger catastrophic tipping elements in the climate system is steadily slipping out of view.

Why does humanity find itself in such an unforgiving situation? Part of the answer is very clearly the elites' naked greed and intransigence. I will say more about this in a moment. But to stop there would be to miss the other moral problems. These problems surface when we notice that many existing energy-intensive activities cannot presently be decarbonized,[9] and that, in any case, renewables are not presently being used to replace fossil fuels: At a global level, renewables are merely adding to the total amount of energy consumed.[10] These two facts together explain the uncomfortable parameters of the Kaya identity. Humanity's total annual emissions are a function of the energy intensity of producing a unit of GDP, the carbon emissions per unit of energy consumed, the number of people consuming energy, and the average annual energy consumption per person. Yet while there have been very significant gains in the first two components of this equation since 1950, these gains in efficiency have been completely swamped by growth in energy consumption per person and growth in global population size.[11]

In short, a continually growing global economy is presently barreling toward 2 degrees of warming and more *without* the ability to rapidly decarbonize long-haul aviation, transoceanic shipping, cement and steel production, most industrial heat processes, the mass production of fertilizer, and so on. In perhaps another ten or fifteen years' time, unless all of these processes suddenly stop, the eventual rise in temperature may trigger irreversible tipping elements in the climate system whose effects could be catastrophic for human beings and countless other species in the long run. The reason we find ourselves in this unforgiving situation is not merely because the elites have been so intransigent, but also because total human energy consumption is growing faster than it has been or even could be decarbonized at present. We therefore cannot think responsibly about the climate crisis without evaluating both energy consumption per capita and global population size. And yet these tasks continue to be systematically avoided in mainstream policy discussions.

Consumption Growth and Population Growth

Nearly two decades ago, Stephen Pacala and Robert Socolow proposed the stabilization wedge approach to mitigation.[12] They outlined fifteen technological changes that, if put in place, would each reduce annual emissions by one billion tons of carbon in fifty years. To give a sense of the scale of the technological changes involved, one of these wedges called for building seven hundred new nuclear power plants, twice the number that has ever been built. A second wedge called for reforesting an area equivalent to twice the arable land of the United States. A third wedge called for installing two million 1-megawatt windmills on an area larger than the United Kingdom. Despite the scale of these projects, the authors noted that on the current emissions trajectory, the world would need to implement eight such wedges simultaneously—not in order to *cut* annual global emissions—but just to keep them from *doubling*.

As Philip Cafaro has noted in his reflections on this topic, none of the proposed stabilization wedges makes a serious effort to limit total human consumption growth.[13] For example, meat-eating contributes approximately two and a half billion tons of carbon per annum to global emissions.[14] In fact, there are so many animals now raised for human consumption that if you were to put them all on a scale, the caged animals would outweigh all of the wild mammals and birds left in the world by a ratio of 10 to 1.[15] Meanwhile, the UN projects that between 2000 and 2050, the worldwide production of animals for human consumption will double, from 60 billion to 120 billion animals produced per annum. This growth will inevitably greatly increase agricultural greenhouse gas emissions. Yet if the world were merely to hold fixed the number of animals produced for human consumption at the current level, this would reduce projected emissions by the equivalent of two and a half of the stabilization wedges that I just mentioned.[16] Indeed, even preventing just half of the expected growth in animal consumption to 2050 would reduce projected carbon emissions by the equivalent of tripling the global number of nuclear power plants.

A similar analysis applies to other extremely emissions-intensive activities, such as aviation. Global air passenger traffic is currently doubling roughly every fifteen years.[17] Preventing half of the projected increase to 2050 would likely reduce emissions by two billion tons of carbon per year or two of the wedges mentioned, whereas holding flights at the current level would likely cut four billion tons annually or four stabilization wedges.[18] Finally, consider the human population, which has doubled from four billion to eight billion people in the fifty years since 1973, and is projected to reach ten billion people by some time in the 2060s. With every additional 500 million people on the planet, global greenhouse gas emissions are expected to grow by at least one billion tons of carbon per annum.[19] Thus, if the human population were to stabilize at the current level of eight billion by 2050, that would reduce projected global emissions by likely four billion tons of carbon annually. However, if global population were to follow the UN's "high growth scenario" and reach twelve billion people by the 2060s, then reforesting an area three times the size of the continent of Australia would still not be enough to outweigh the growth in carbon emissions.[20]

In short, human consumption growth and global population size will both make an enormous difference to how much carbon is emitted and therefore whether catastrophic climate tipping points are triggered. Yet limits on these variables are rarely even mentioned in technical climate policy discussions.[21] Cafaro, who has tried to draw attention to this problem, has asked philosophers to weigh in on the question of why these topics continue to be systematically avoided. I would now like to take the opportunity to offer my response.

As I see it, there are two principal reasons why many experts neglect to discuss limits on total consumption growth. The first reason is that most existing analyses of consumption growth come from economic models, and economists have been incredibly cavalier about the risks of climate change.[22] Consider for instance that in the standard DICE model, it is strictly impossible to crash the global economy even with six or more degrees of global warming.[23] At six degrees of warming, the daily temperature in New York would be deadly to mammals for likely two months of the year; most existing insect species would be extinct; and with enough time for the melting to take place, crocodiles would once again be living in what is now the Arctic.[24] Yet in the DICE model it is impossible for changes even of this magnitude to crash the sum of human welfare. That is because, among other things, GDP is taken as a suitable proxy for welfare and it's assumed that nothing that happens outside can affect GDP in our predominantly indoors service-sector economy. Here is a statement of this mainstream idea from the Intergovernmental Panel on Climate Change: "Economic activities such as agriculture, forestry, fisheries, and mining are exposed to the weather and thus vulnerable to climate change. Other economic activities, such as manufacturing and services, largely take place in controlled environments and are not really exposed to climate change."[25]

I mention this Nobel Prize–winning idea only to set it aside. The other reason why analysts tend not to discuss constraints on total consumption growth is because the standard optimal policy framework is sum-total utilitarian.[26] This is clear from the maximands of the integrated assessment models used to evaluate alternative consumption and emissions pathways. These models tell us to maximize an additive intergenerational social welfare function that takes as inputs estimates of all present and future preferences over consumption. Yet there is no way for such an approach to provide morally sound guidance concerning what should be done about human consumption growth. On the contrary, the approach produces morally abhorrent results, and the practitioners have long understood this fact.

To see this, consider first the question of optimal population size.[27] One might think that with the sorts of impacts that I have outlined, a utilitarian approach to climate policy would recommend attempting to stabilize or even gradually reduce the size of the global population. In fact, a sum-total utilitarian approach makes the opposite recommendation: It tells us to adopt policies that will grow the population. The reason for this recommendation, called the Repugnant Conclusion by philosophers of population ethics, is the fact that sum-total utilitarianism fundamentally cares only about the sum of welfare and not about the fate of anyone in particular. Accordingly, it will direct us to take steps to grow the population, to many tens of billions of people, all of whom consequently have lives that are barely worth living, just as long as the total amount of welfare summed across persons is larger (in much the way that a million bottles of wine, each containing a single drop, will hold more wine in sum than a single full bottle).[28]

This problem is just a special instance of a more general problem with the utilitarian approach to climate policy. Another instance arises when we try to use the approach to tell us how much present-day consumption we should be prepared to sacrifice for the sake of future generations. Here is the second issue in a nutshell. For every resource that we can either consume now or withhold from consumption and invest for the future, the objective of maximizing an additive intergenerational social welfare function will demand that we *deny* ourselves the use of this resource and invest it for the future, just as long as this would produce positive net returns to aggregate welfare summed across an indefinite series of future generations.[29] Applied to emissions reductions, this approach will therefore tell us to literally starve the present generation if, say, a drastically lower population would allow the biosphere to recover from such ongoing processes as the sixth extinction and thereby enable even higher consumption paths in the more distant future. In other words, as long as drastic cuts in present-day consumption are technically feasible and would maximize total welfare summed across many generations, an intergenerational utilitarian social objective will require the present generation to undertake the severest possible cuts in human energy and land use, irrespective of the potentially unbearable sacrifices thereby implied for people living in the present.

Even the practitioners of this approach find these conclusions unacceptable. To avoid them, the conventional optimal policy framework has come to make use of two ad hoc devices. The first is the device of discounting the welfare of future generations in order to make room for more present-day consumption. The second is the tactic of ignoring global population size by treating it as exogenous to the analysis. For obvious reasons, these tactics make the conventional approach useless for evaluating total human consumption growth. Yet both tactics are widely used and indeed felt to be absolutely necessary. Otherwise, it is feared, the present generation will literally have to "starve itself to benefit future generations," while optimal climate policy will call for growing rather than stabilizing the future human population.[30]

These fears rest on a mistake, however. The conclusions that climate economists seek to avoid follow only if we, the present generation, have a duty to maximize the sum of preference satisfaction across all of time. Yet this assumption misdescribes our duties of intergenerational justice in two important ways. First, we do not owe it to future people to help them get more of whatever they happen to want. The desires of future people for robotic maids and vacations in space are less urgent than the needs of poor people today for clean water, electricity, and basic sanitation facilities. Conversely, irreversible losses in the future of arable land and other life-sustaining resources cannot be offset by the ability of today's wealthy consumers to continue their frequent air travel to distant locales. By treating present and future "preferences" for these diverse goods as if they were all on a moral par, and telling us to maximize the extent of preference satisfaction across all of time, the utilitarian approach to climate policy misrepresents what we owe to our contemporaries as well as to others more distant in time.

Second, we do not owe it to future people to bring more and more of them into existence as a means of maximizing the sum of welfare. In fact, we do not owe it to any of our countless, merely possible offspring to bring any of them into existence at all. Instead, what is true is that we *will* owe various things to everyone who ever comes to exist, if and when they come to exist—things ranging from police protection, the rule of law, and a system of functional educational institutions, to all of the other rights and entitlements that all living persons should be understood to have as a matter of social and international justice. The real problem of intergenerational justice, then, is not how we can avoid starving the living or ballooning the population in the name of creating the maximum amount of welfare in the future. These problems are artifacts of a consequentialist utilitarian approach that is entirely misguided in a case like climate change. Instead, the real problem of intergenerational justice is how to understand our present-day duty to preserve rights-respecting institutions for the indefinite future, and in particular how to conceive this duty so that it will be appropriately demanding. I will now describe how I think this challenge is best understood and how I think it can be solved.

An Approach to Intergenerational Justice

In my view, the most significant philosophical obstacle to defensible moral reasoning concerning the requirements of intergenerational justice in the climate change context comes not from the Repugnant Conclusion but from the Non-Identity Problem. This is because, even if our failure to cut global emissions quickly will have truly catastrophic consequences in the future, the Non-Identity Problem makes it difficult to see why the present generation will be wronging anyone by continuing with business as usual. The difficulty arises from three observations, not always clearly appreciated.

First, if what I have said in this chapter is even remotely true, it will be immensely costly to cut global greenhouse emissions to net zero before they threaten to trigger irreversible climate tipping points. After all, there are billions of people around the world who still lack electricity, adequate nutrition, access to medicine, and basic sanitation facilities, not to mention all of the other things that you and I take for granted. And yet at present, the global economy cannot operate even at current levels of output without using enormous quantities of fossil fuels for many decades. Alternatives to many combustion processes, petrochemicals, steel, concrete, and other carbon-intensive materials have yet to be invented, and their mass production is likely to take decades even in the best of circumstances. Accordingly, if the world is to avoid temperatures that threaten catastrophic long-run dangers, then at least a temporary halt in the staggering rate of total human consumption growth appears to be necessary in the short term. Yet for this purpose, our governments would need a very good justification.

Two further observations make this justification elusive on a nonconsequentialist, person-regarding view of intergenerational justice. The first is that by continuing with business as usual, we will not actually be undermining the possibility of just institutions in the future. For even when the environment has been despoiled in, say, a couple of hundred years, it will continue to be possible for institutions to fairly ration the remaining resources and treat all living people in what will then be the morally required way.[31] It's just that, because the environment will have been ruined, maintaining just institutions in the future will at best afford everyone a much lower quality of life than some of us are able to enjoy today.

At the same time, it does not appear that we will be wronging anyone by lowering the standard of living attainable in the future to a very low level. For the present generation is in fact benefited by not having to make large sacrifices for the sake of the distant future; and even if it were to make such sacrifices, this would not actually *help* the people who will see the world's coastal cities swallowed by the oceans several centuries from now. On the contrary, these particular people would end up never being born at all. After all, rapidly reducing global greenhouse emissions would have countless effects on what people today are able to do and be, and consequently also on whom they ever meet, and with whom they decide to have kids. But just as you would never have been born had your mother never met your father and instead had children with some other individual, so, too, the persons who will live if we continue with business as usual will not be one and the same as the persons who will be born if the world makes rapid and extensive changes to its land-use practices, all of its transportation systems, and all of its energy infrastructure. Like ripples in a pool, the behavioral changes that will then inevitably follow will alter not only who meets whom and has children with them in the present generation, but also whose children will go on to have children of their own. Thus, if emissions are rapidly reduced today to avoid triggering catastrophic climate tipping points, this will not actually help the people who, several centuries from now, would otherwise live through the loss of the world's cities from dozens of feet of sea level rise. Instead, because of the countless behavioral changes that will then inevitably follow, these particular people will end up never being born at all.

In short, from the perspective of a nonconsequentialist and person-regarding view of intergenerational justice, it's not clear why the present generation has a duty to rapidly reduce its consumption growth *even if* this is necessary to avoid triggering climate tipping points that will eventually prove catastrophic. For it appears that the people who will live through such catastrophes will not be benefited if we do, whereas the duty not to undermine the possibility of just institutions does not seem to rule out continuing with business as usual either. So from a non-utilitarian and person-regarding point of view, how should we reason about intergenerational justice so that the implications for what the present generation is morally required to do will be both illuminating and suitably demanding?

In thinking about this question, I have come to the conclusion that we should understand our duties of intergenerational justice as owed first and foremost to our younger contemporaries. In other words, I do not think that we should try to dissolve the fact of non-identity, much less to ignore it. It is simply true that, no matter what we do, our climate policies will affect who and how many people will come after us. If we continue on our current course, then many children will one day be born in makeshift environmental refugee camps, to parents who tragically met there for the first time in part because of our inaction. Yet as long as the lives of these children will still have been worth living, they would not be helped by our changing the course of world history; instead, they would end up never existing. The task for moral theory as I now see it is, not to deny this simple truth, but to make it strictly *irrelevant* to the content of appropriately demanding present-day environmental duties. Moreover, I believe that three ideas will prove sufficient for this purpose.

First, it's false that climate disasters attributable to today's emissions will befall *only* people who will be born in the future. On the contrary, some children who are already alive will one day be gravely harmed by the emissions that the global economy will put out tomorrow. We have no way of knowing who these particular children are. But even if the lag between emissions and destructive environmental consequences is decades long, we can be certain that some such children are already alive and that they will eventually be gravely harmed. Second, we need to recognize that as a matter of justice, all living persons have claims of varying degrees of moral importance, not just a single claim to "preference satisfaction"; and that institutions must be erected that continuously give every living person's diverse claims the appropriate weight as soon as and for as long as they live. But these two ideas together already show that, by failing to cut much of the world's present-day luxury consumption, the older generation in charge today in all of the world's countries is wronging some of the children born around the world yesterday. For it will create a situation in which, later in life, not all of these children will be able to escape conflict, famine, and drought, for no reason more important than the present-day consumption of luxury experiences and goods. Letting this happen *now* is incompatible with the duty to ensure, as far as we can, that just institutions are in place and at all times give the appropriate weight to every living person's diverse moral claims—some urgent and some much less so.

These reflections are, I think, enough to show that requirements of intergenerational justice in respect of climate change can be coherently worked out over periods of time approximating a single human lifespan. The duty to ensure, as far as we can, that just institutions are continuously maintained for whoever exists, as soon as and as long as they exist, already imposes on us the appropriately demanding requirements over this particular time horizon: The content of these requirements is given by our best theories of social and international justice. So, for instance, if our theories tell us that everyone's right to avoid unnecessary famine is more urgent or basic than anyone's claim to eat grass-fed beef or travel long distances by plane for vacation, then we already know what we are morally required to do as a matter of intergenerational justice: Dramatically restrict emissions from animal agriculture, and dramatically restrict aviation emissions from tourism until a truly carbon-neutral alternative to jet fuel is invented. Otherwise, we will be wrongly indulging less important interests in luxury consumption at the expense of survival interests that are comparatively much more important.

However, the ideas that I have outlined can also be used to generate requirements of intergenerational justice across all of history: in particular, requirements on *each* generation in history to show appropriate environmental concern *at* each stage in history. Indeed, I believe that these ideas can even be used to reason about when and what form of constraints on population growth are morally appropriate, and to do so in a way that remains sound even when the number and identity of future people will be affected. What we need to do for this purpose is to recognize a third moral principle: Unless there is an adequate justification, it wrongs people to create a situation in which it will no longer be possible to avoid either wronging or harming them in the future.

To articulate the reasoning warranted by this principle, I invite you to consider the following stripped-down example. Imagine a small-scale society consisting of a mother and a father and two children. The rules that the parents impose on everyone play the role of the society's basic political and economic structure. Meanwhile the land has been overfarmed for many generations. As a result, the family now lives at the edge of their means, in the following sense: If the parents were to have additional children, then it would no longer be possible for everyone to be adequately fed. The question is why it would be wrong for the parents to allow this to happen, assuming that the additional child would still have a life that is worth living and doing otherwise would mean that the additional child would end up never existing. And the answer is that the parents would thereby wrong one another and their existing children.

Here is the reasoning that leads to this conclusion. 1. As a matter of justice, persons have claims to the protection of diverse goods and interests, and these claims have varying degrees of moral importance. Therefore, the rules of a society ought to avoid indulging claims to less important interests at the expense of interests that are in the circumstances even more important. Yet this is precisely what the parents would be doing by failing to recognize a moderate restriction on their procreative freedoms—"moderate," because they have already allowed themselves to have two children. By failing to recognize a moderate restriction on their procreative freedoms, the parents would create a situation in which an even more serious restriction—on a right that normally protects an even more urgent interest—would be needed to meet the diverse claims of everyone in existence. For then there would be five mouths to feed, each with identical claims on sustenance, and so even access to a bare minimum of *food* would now have to be strictly rationed, on pain of committing a grave injustice.

2. In permitting this situation to arise, the parents would be wronging everyone alive even *before* the arrival of their third child. This is because of what I have called the principle of disjunctive wrongs: In the absence of an adequate justification, it wrongs people to act in ways that will necessitate either wronging or harming them in the future. But that is exactly what would happen if the parents allowed themselves to have additional children in their already dire circumstances. The parents would need to harm everyone already alive by imposing a painful system of food rationing, or else wrong everyone by letting someone starve and thereby committing a grave injustice. In creating this situation, moreover, they would be acting without adequate justification, because they would be privileging less important interests over interests that in the circumstances were even more important. Therefore, the parents would be acting wrongly—in particular, wronging their existing children—and it is strictly irrelevant that if they were to act *rightly*, the identity and size of the population would also be affected.

In short, there is a nonconsequentialist way of reasoning about the demands of intergenerational justice that is untouched by the non-identity issue. Moreover, this structure of reasoning can be extended indefinitely backward and forward in time. To see this, notice that the parents in our story will themselves have been wronged by *their* parents if, in order to fuel the grandparents' runaway luxury consumption, the grandparents avoidably contributed to overfarming and thereby reduced the earth's carrying capacity. That prior wrong will have had the identical structure: Failing to put in place restrictions on less important, luxury consumption-related freedoms, and thereby creating a situation in which an even more serious restriction—on much more important, procreative freedom-related interests—would be needed to respect the diverse claims of everyone already in existence at the time that the grandparents were disastrously overfarming.

This, then, is what I have come to think is the right way to reason about requirements of intergenerational justice both in general and in the context of the climate crisis. The reasoning illustrated is able to establish an unending chain of person-regarding nonutilitarian intergenerational obligations that is indifferent to the identity of future generations. It is sometimes said that thinking about intergenerational ethics exclusively in terms of what we owe to our younger contemporaries ignores the independent claims of as-yet unborn future generations. But as I have just shown, that is a mistake. It is an essential part of the explanation for wrongdoing that I have given that future people *will* have moral claims on everyone as soon as they come into existence and that—unless decisive steps are taken in the present—even greater sacrifices will then be morally necessary to honor these rights, on pain of grave injustice.

Indeed, what the reasoning shows is that at *every* stage in history, people have strong justice-related reasons to put in place all those restrictions on existing rights and freedoms that, unless they are put in place, will eventually require even more serious sacrifices having to be made by their contemporaries—lest their contemporaries fall afoul of the very same moral imperative in their relations with *their* future brethren. In turn, the essence of intergenerational *in*justice is a form of "kicking the can down the road": *failing* to put in place restrictions on existing rights and freedoms that are necessary to prevent even more serious sacrifices having to be made later by one's younger contemporaries for the purpose of enacting justice and securing every living person's diverse moral claims to be free from avoidable famine, drought, resource wars, and so on.

As it turns out, this is exactly what today's most powerful people are unconscionably doing now. And it is exactly how you and I were wronged by the people who were in charge several decades ago. They failed to make changes to energy systems that, though they would certainly have been costly and socially disruptive, would have been much less burdensome to undertake in the time that was then still available to prevent catastrophic climate change. Instead, they chose to indulge much less urgent interests in consumption and accumulation and left us in a much more dire situation. By kicking the can down the road, they thereby wronged the members of our generations. In turn, today's power-holders are doing the very same thing to the children born around the world yesterday. They are indulging various less important interests in accumulation and consumption rather than reducing luxury emissions and accelerating the global energy transition, ensuring that even more burdensome steps will later be morally required—of these children, for this purpose—else the world eventually collapse from catastrophic climate change.

These are the sorts of general conclusions that are rendered sound by the form of non-utilitarian moral reasoning that I have set out. However, articulating a coherent approach to reasoning about a subject does not by itself answer urgent practical questions. In passing, I have suggested that governments have strong reasons of intergenerational justice to put in place at least temporary restrictions on industrial animal agriculture and luxury aviation emissions. Yet I have not reached any conclusions even in passing concerning the permissibility of any particular means of influencing global population size. Arriving at considered conclusions about even indirect and noncoercive ways of doing this requires thinking through difficult questions concerning the relative importance of the interests at stake. To understand what precisely we morally ought to do about the climate crisis and how quickly—which activities we should enable, encourage, and restrict—there is no alternative in my view than to think through these questions one by one, using an approach that correctly identifies the central imperative of intergenerational injustice.[32]

Nature, Uncertainty, and Future Wrongdoing

Unfortunately, this will still not be enough. For there are dimensions that are absolutely central to the ethics of the climate crisis but that, in spite of many attempts to think them through, I hardly know where to begin, and I do not think I am alone. One such absolutely central issue is how to think about the impact of our actions on all of the other animals and the natural world. These impacts are so enormous that, like others, I often find myself paralyzed and recoil from learning more.[33] Already there are forest fires in the world that burn billions of fellow creatures in a single unstoppable inferno. In the summer of 2023, the ocean temperature breached 100 degrees Fahrenheit in a place tragically named Manatee Bay off the coast of southern Florida. Such temperatures represent invisible forest fires under the surface of the sea, decimating kelp forests that have sustained entire ancient ecosystems. As temperatures rise, destruction on an even larger scale will occur all around the world. At the same time, other less complex organisms will flourish in radically transformed environments. How should we incorporate such facts into our moral judgments? How much should we be prepared to sacrifice in order to arrest global warming even faster? For me, the temptation is to privilege the existing so-called higher animals in my reasoning, and to adopt an individual animal-regarding approach, modeled on the non-utilitarian person-regarding one above. Yet even though the same population ethics issues apply to animals with full force, here I have very little confidence that I have found anything like the right approach.[34]

At the same time, I am convinced that the conventional policy framework represents an enormous moral crime against the other animals. This framework generally values animals only instrumentally for their "ecosystem services," and more rarely because *people* feel distress at the thought of countless animals burning. When the conventional approach does give independent weight to animal suffering, it produces counterintuitive conclusions in the climate change context. For example, it suggests paving over the jungles if the lives of the innumerable animals there are (plausibly) filled with suffering,[35] and that industrial agriculture should privilege greenhouse gas–intensive grass-fed beef because of the greater sum of pain felt by factory-farmed chickens.[36] Still, I hesitate to press such observations into criticisms because, as I have said, I have very little sense of the shape of a reasonable philosophical alternative.

My sense of being at sea in the ethics of the climate crisis is heightened by another dimension that is central to responsible moral reasoning in this context. This is the problem of how to make fateful social decisions under conditions of profound uncertainty about the consequence of these actions. In this chapter, I have stressed that irreversible climate tipping points may well be triggered at nearby temperatures, and that even in the best of circumstances these nearby temperatures will be enormously costly to avoid. Yet although we know that certain tipping points could well be triggered even by the presently locked-in global temperature, for various tipping points there is no scientific basis for determining whether they are likely to be triggered in twenty years' time, fifty years' time, or in a century from now. In conventional integrated assessment modeling, this problem of deep or Knightian uncertainty is "solved" by *stipulating* probability distributions that represent all known catastrophic tipping points as unlikely tail events.[37]

In one way, this tactic for dealing with uncertainty is unsurprising: Probability distributions are required to calculate what would maximize expected welfare. Still, there is no scientific basis for most of the implied empirical judgments. For instance, some of the worst effects of climate change will occur when countries like India, Pakistan, and China go to war over migrant flows or dwindling water. Yet no self-respecting social scientist would agree to construct even rough probability distributions for such world-historical social tipping points. The result is that all existing calculations of the carbon price that would allegedly maximize expected global welfare are based on willful ignorance and unfounded stipulation.

This brings me to a final moral dimension of the climate crisis whose difficulty I believe is still not fully appreciated. The problem is how we, you and I, should reason about our moral obligations once we admit to ourselves that the powerful will predictably not obey any of the most important imperatives of climate ethics. Discussions of the climate crisis all-too-frequently ignore this problem by asking questions exclusively from the undivided perspective of "society," or alternatively from the perspective of a morally well-motivated social planner. Yet both of these assumptions depart dramatically from reality. Just consider all of the new fossil fuel projects that are currently being licensed. The past few years have seen record worldwide coal demand; 195 new coal plants are currently under construction.[38] The United Kingdom has recently issued hundreds of new oil drilling blocks in the North Sea.[39] The United States has engaged in record oil and gas exploration and now produces more than 12 million barrels of oil per day.[40] Germany is building several new liquified natural gas terminals; once the enormous liquification and transport energy costs are counted, these terminals will be responsible for more emissions than simply digging up and burning German coal.[41] Meanwhile, India and China plan to complete dozens of new oil, coal, and gas projects by 2025.[42] Uganda has hundreds of new oil wells under construction by the Chinese.[43] In Namibia, billions of barrels of oil were recently discovered and are now under development.[44] In South Africa, trillions of cubic feet of natural gas have been discovered and the fossil fuel companies are eagerly lining up to pump them.[45] In Canada, the tar sands of Alberta, which contain the dirtiest crude on earth,[46] currently produce three million barrels of oil a day, still far from the maximum theoretical yield. However, the capital projects that have been committed in the tar sands through 2025 should be enough to breach 2 degrees of global warming all by themselves.[47]

How are we to reason about climate ethics once we admit that these reckless actions will not stop? The first step is to understand that there will be auto-catalytic social effects. Past inaction on emissions will produce major social disruption from storms, droughts, heat waves, and crop failures. In turn, these destructive consequences will provide the occasion for yet more wrongdoing by powerful people bent on exploiting all available fossil fuels. Effects of this kind are on the horizon everywhere we care to look. For example, a large number of US houses are now significantly mispriced because of increasingly uninsurable flood risk.[48] When these losses are realized, the impact may well set off another financial crisis. In other words, the housing market may yet present a second historic opportunity to socialize the losses to which finance capital has become exposed, while shifting the costs of rebuilding to an increasingly resentful working class living in coastal municipalities with a permanently lower tax base.

The next step in thinking clearly under such predictably non-ideal conditions is to recognize that auto-catalytic social effects can generate acute moral problems for the nearby actors. To see this, consider proposals to increase funding for research into solar geoengineering. In coming years, societies will increasingly face what can be thought of as the air conditioner problem. As temperatures rise, hundreds of millions of additional people will need air conditioning merely to survive. Developing the electrical capacity for this purpose will then compete directly with the aim of accelerating the energy transition, since yet more energy infrastructure will need to be built out and decarbonized. In fact, this problem is merely an instance of a much bigger looming mitigation-adaptation tradeoff. As climate change damages countless buildings, washes out roads and bridges, and levels coastal cities, the energy, labor, and material costs of rebuilding all of this destroyed infrastructure will likewise come on top of all that will be needed to transform the global energy system.

In this context, solar geoengineering could in principle make the adaptation-mitigation tradeoff less severe, assuming that it does not turn out to be climatologically counterproductive. This is, of course, a very tall assumption. Solar geoengineering would do nothing to prevent increasingly dangerous ocean acidification from ongoing carbon emissions; it could well shift precipitation patterns that provide the water supply for hundreds of millions of people in politically unstable areas; and the sudden termination of a scaled-up geoengineering effort during an extended global conflict could be apocalyptic for the entire human species.[49] However, to appreciate why the moral challenge is not limited to these issues, suppose that solar geoengineering would not have these climatologically counterproductive consequences. Then in principle its use could make the adaptation-mitigation tradeoff less severe by reducing the amount of damaged infrastructure that would need to be rebuilt at high energy, labor, and material cost. The savings in human life and treasure could then be used to accelerate the global energy transition.

But of course, these real savings could just as well be used to expand the production of fossil fuels, unsustainable consumption growth, and greenhouse gas–intensive land use. In fact, some US lawmakers have already proposed using geoengineering for this purpose "instead of forcing unworkable and costly government mandates on the American people."[50] And a similar politics now governs the various technologies of carbon capture. In the United States, there is already an extensive network of pipelines for transporting carbon dioxide removed from the air. The pipelines are used overwhelmingly not to sequester carbon but to enable enhanced oil recovery in fracked wells. Meanwhile, no government on the planet has developed carbon capture capacity to the fantastical levels that conventional climate modeling takes for granted,[51] yet numerous governments have been busy extending generous subsidies for enhanced oil recovery. Against this background, it is no wonder that fossil fuel executives have confidently told investors that carbon capture technology will keep their industry in business for at least the remainder of the century.[52]

In these unpropitious circumstances, the danger facing anyone who stands to affect the development of technologies such as carbon capture and solar geoengineering is the risk of creating a situation in which—thanks to these scientific efforts—various powerful actors are enabled to deepen the intergenerational injustice that is at the heart of the climate crisis. Instead of deploying the real savings that might be enabled by these technologies to accelerate the global energy transition, the economic elites who have time and again shown themselves to be indifferent to considerations of justice may well use these savings to develop yet more fossil fuels and unsustainable land use—and thereby create a situation in which even more serious sacrifices will be necessary by our youngest contemporaries in order to forestall future catastrophic climate change.

This prospect raises an urgent practical question for anyone who finds themselves in a similar position. How should wrongdoing that will be a predictable consequence of our actions affect how we understand our moral, political, and professional obligations? Is the injustice that others will commit entirely on them? Or does appropriate concern for justice demand that we change what we are doing, too? In thinking about the ethics of the climate crisis, it is difficult to overstate the practical significance of these questions.[53] For the worst effects of climate change will come not from the storms or the fires or the simultaneous crop failures. The worst effects will come when the economic elites and other people respond morally wrongly to the burgeoning human fallout. In addition to the other philosophical challenges that I have noted, this fact makes understanding what you and I ought to do about the climate crisis among the hardest moral problems.

Notes

1 Eric Posner and David Weisbach, *Climate Change Justice* (Princeton, NJ: Princeton University Press, 2010), 86. For critical discussion, see Simon Caney, "Two Kinds of Climate Justice: Avoiding Harm and Sharing Burdens," *Journal of Political Philosophy* 22 (2014): 124–149, and Stephen Gardiner, "The Threat of Intergenerational Extortion: On the Temptation to Become the Climate Mafia, Masquerading as an Intergenerational Robin Hood," *Canadian Journal of Philosophy* 47 (2017): 368–394.

2 For a review of the scientific literature on tipping elements in the climate system, see Seaver Wang, Adrianna Foster, Elizabeth A. Lenz et al., "Mechanisms and Impacts of Earth System Tipping Elements," *Reviews of Geophysics* 61 (2023): 1–81.

3 Chris A. Boulton, Timothy M. Lenton, and Niklas Boers, "Pronounced Loss of Amazon Rainforest Resilience since the Early 2000s," *Nature Climate Change* 12 (2022): 217–278.

4 Peter Ditlevsen and Susanne Ditlevsen, "Warning of Forthcoming Collapse of the Atlantic Meridional Overturning Circulation," *Nature Communications* 14 (2023): 4254.

5 To appreciate the scale of the predicted changes, note that "the yearly averaged surface temperature change exceeds 1°C *per decade* over a broad region in northwestern Europe, and for several European cities, temperatures are found to drop *by 5° to 15°C.*" See René M. Westen, Michael Kliphus, and Henk A. Dijkstra, "Physics-Based Early Warning Signal Shows that AMOC Is on Tipping Course," *Science Advances* 10 (2024): 1–11, at 3 (my emphasis). See also Stefan Rahmstorf, "Is the Atlantic Overturning Circulation Approaching a Tipping Point?" *Oceanography* 37, no. 4 (2024): 16–29, at 27: "A full shutdown of the AMOC would have truly devastating consequences for humanity and many marine and land ecosystems. . . . The cold air temperatures then expand to cover Iceland, Britain, and Scandinavia. The temperature contrast between northern and southern Europe increases by a massive 4°C, likely with major impact on weather, such as unprecedented storms. As we have seen in the paleoclimate data for Heinrich events, major precipitation shifts in the tropics would likely cause drought problems in the northern tropics of America as well as Asia. Seasonal changes will be even larger than these annual mean changes. Other simulations predict a significant increase in winter storms in Europe and a strong reduction of crop yield and pasture there. . . . The IPCC summarized the impacts: 'If an AMOC collapse were to occur, it would very likely cause abrupt shifts in the regional weather patterns and water cycle, such as a southward shift in the tropical rain belt, and could result in weakening of the African and Asian monsoons, strengthening of Southern Hemisphere monsoons, and drying in Europe.' Some further consequences include major additional sea level rise especially along the American Atlantic coast, reduced ocean carbon dioxide uptake, greatly reduced oxygen supply to the deep ocean, and likely ecosystem collapse in the northern Atlantic."

6 The updated carbon budget figures in the next two paragraphs are drawn from Piers M. Forster, Christopher J. Smith, Tristram Walsh et al., "Indicators of Global Climate Change 2022: Annual Update of Large-Scale Indicators of the State of the Climate System and the Human Influence," *Earth System Science Data* 15, no. 6 (2023): 2295–2227. In reviewing these data, I have also benefited from a discussion by Zeke Hausfather. Unlike conventional IPCC estimates, the carbon budgets reported by Forster and co-authors assume that carbon capture technology will not be deployed on a massive scale later this century to rapidly reduce the stock of atmospheric greenhouse gases by tens of billions of tons of carbon. I indicate why I think this assumption is appropriate in the final section of this chapter.

7 See James E. Hansen, Makiko Sato, Leon Simons et al., "Global Warming in the Pipeline," *Oxford Open Climate Change* 3, no. 1 (2023): 1–33.

8 Paul Voosen, "Ship Fuel Rules Have Altered Clouds and Warmed Waters: An Unforeseen Test of Reverse Geoengineering Unfolds Above the Oceans," *Science* 381 (2023): 467–468.

9 Among others, these activities include flying more than 1,000 miles (for example, from continent to continent), shipping heavy commodities across the oceans, most existing cement and steel production, the fuel sources for many industrial heat processes, the energy-intensive production of fertilizer feedstock and its greenhouse gas–producing application to soils. In some cases, such as long-haul aviation, the fossil fuel alternatives have yet to be invented. In other cases, alternatives exist but their mass production will take many decades. Recall that barely 1% of the global vehicle fleet is electric, and already the demand for lithium and cobalt is stretching global mining capacity: see Chengjian Xu, Qiang Dai, Linda Gaines et al., "Future Material Demand for Automotive Lithium-based Batteries," *Communications Materials* 1 (2020): 1–10, at 4: "Given the magnitude of the battery material demand growth across all scenarios, global production capacity for Li, Co, and Ni will have to increase drastically. For Li and Co, demand could outgrow current production capacities even before 2025."

10 See Hannah Ritchie, Max Roser, and Pablo Rosado, "Energy," *Our World in Data* (2022), ourworldindata.org.

11 "GDP/capita and population growth were the main drivers of the increase in global emissions during the last three decades of the 20th century. . . . At the global scale, declining carbon and energy intensities have been unable to offset income effects and population growth and, consequently, carbon emission have risen." Intergovernmental Panel on Climate Change (IPCC), *Climate Change 2007: Mitigation*, ed. Bert Metz, Ogunlade Davidson, Peter Bosch, Rutu Dave, and Leo Meyer (2007), 107. For further discussion see Philip Cafaro, "Beyond Business as Usual: Alternative Wedges to Avoid Catastrophic Climate Change and Create Sustainable Societies," in *The Ethics of Global Climate Change*, ed. D. Arnold (Cambridge: Cambridge University Press, 2011): 192–215.

12 See Stephen Pacala and Robert Socolow, "Stabilization Wedges: Solving the Climate Problem for the Next 50 Years with Current Technologies," *Science* 305 (2004): 968–972.

13 See Cafaro, "Beyond Business as Usual." In the next two paragraphs, I reproduce some of the analysis that Cafaro sets out on pp. 201–208.

14 United Nations Food and Agriculture Organization, *Livestock's Long Shadow: Environmental Issues and Options* (Rome: 2006), 112, discussed in Cafaro, "Beyond Business as Usual," at 202.

15 Yinon M. Bar-On, Rob Phillips, and Ron Milo, "The Biomass Distribution on Earth," *Proceedings of the National Academy of Sciences* 115 (2018): 6506–6511.

16 Gidon Eshel and Pamela Martin, "Diet, Energy, and Global Warming," *Earth Interactions* 10 (2006), Paper No. 9, discussed in Cafaro, "Beyond Business as Usual," at 202.

17 Intergovernmental Panel on Climate Change (IPCC), *Climate Change 2007: Mitigation*, 334.

18 Cafaro, "Beyond Business as Usual," 203.

19 This is assuming the present global per capita average emissions of two tons of carbon per person per annum. It's important to note that, in a world in which the average person is still absolutely poor, the present global per capita average emissions of two tons of carbon per annum is less than 15% of the average American's annual emissions. For this reason, the figure is not merely unrealistically low but would likely also amount to an injustice once additional people come into existence but before most industrial processes can be decarbonized.

20 See Cafaro, "Beyond Business as Usual," 206–208 for his corresponding discussion. I have added the Australia figure, using Pacala's and Socolow's estimate that 300 million hectares of new tree plantations would keep one billion tons of carbon annually from entering and remaining in the atmosphere in fifty years. If global population were to grow from eight billion people to twelve billion people later this century, as projected in the UN's "high growth scenario," and if the average new person were to emit the current 2-ton global per capita average, then the required offset would be eight billion tons of carbon per annum, or eight times the amount removed by 300 million hectares of mature trees using Pacala's and Socolow's estimates. However, there are fewer than 800 million hectares of land on the entire continent of Australia. Thus, offsetting the emissions of an additional four billion people by reforesting 2400 million hectares (again using Pacala's and Socolow's stabilization wedge estimates) would require an area more than three times the continent of Australia.

21 To be clear, I do not say that the variables themselves are rarely seriously discussed, which is false. (For an overview of some of the philosophical literature, see, e.g., Hilary Greaves, "Population Axiology," *Philosophy Compass* 12, no. 11 (2017): 1–15.) What is rarely mentioned in climate policy discussions is the political morality of placing limits on these variables, or the morality of instituting direct or indirect constraints on

consumption growth and population size. For references to a small but important philosophical literature ignored in mainstream climate policy discussions, see Henrik Andersson, Eric Brandstedt, and Olle Torpman, "Review Article: The Ethics of Population Policies," *Critical Review of International Social and Political Philosophy* 27 (2024): 635–58. See also Colin Hickey, Travis N. Rieder, and Jake Earl, "Population Engineering and the Fight against Climate Change," *Social Theory & Practice* 42, no. 4 (2016): 845–870.

22 For a complementary discussion, see Douglas Kysar's chapter, "Ways Not to Think About Climate Change," in this volume.

23 For discussion of the observations in this paragraph, see Steve Keen, "The Appallingly Bad Neoclassical Economics of Climate Change," *Globalizations* 18, no. 7 (2020): 1149–1177, and Steve Keen, Timothy M. Lenton, Antoine Godin et al., "Economists' Erroneous Estimates of Damages from Climate Change," *Proceedings of the Royal Society A* (2021): 1–35.

24 See C. Mora, B. Dousset, I. R. Caldwell et al., "Global Risk of Deadly Heat," *Nature Climate Change* 7 (2017), 501–506, fig. 4 at 504; and R. Warren, J. Price, E. Graham et al., "The Projected Effect on Insects, Vertebrates, and Plants of Limiting Global Warming to 1.5 C rather than 2 C," *Science* 360 (2018): 791–795, fig. 4 at 792; both cited in Keen, Lenton, Godin et al., "Economists' Erroneous Estimates of Damages from Climate Change," 6–7.

25 D. J. Arent, R.S.J. Tol, E. Faust et al., "Key Economic Sectors and Services," in *Climate Change 2014: Impacts, Adaptation, and Vulnerability. Part A: Global and Sectoral Aspects. Contribution of Working Group II to the Fifth Assessment Report of the Intergovernmental Panel on Climate Change*, ed. C. B. Field, V. R. Barros, D. J. Dokken et al. (Cambridge: Cambridge University Press, 2014), 659–708, at 688. For discussion see Keen, "The Appallingly Bad Neoclassical Economics of Climate Change," 3–4.

26 See for example William D. Nordhaus, "Revisiting the Social Cost of Carbon," *Proceedings of the National Academy of Sciences* 114 (2017): 1518–1523, which describes the influential Dynamic Integrated model of Climate and the Economy (DICE model).

27 In the rest of this section and the next, I draw on and extend formulations of ideas that I first presented in "How Quickly Should the World Reduce Its Greenhouse Gas Emissions? Climate Change and the Structure of Intergenerational Justice," in *Climate Change and Philosophy*, ed. Mark Budolfson, Tristram McPherson, and David Plunkett (Oxford: Oxford University Press, 2021), 301–320.

28 For the classic statement of this implication in axiological form, see Derek Parfit, "Overpopulation and the Quality of Life," in *The Repugnant*

Conclusion: Essays on Population Ethics, ed. Jesper Ryberg and Torbjörn Tännsjö (Boston: Kluwer, 2004).

29 Or, as Kenneth Arrow once put it, "strictly speaking, we cannot say that the first generation should sacrifice everything, if marginal utility approaches infinity as consumption approaches zero. But we can say that given any investment, short of the entire income, a still greater investment would be preferred." Kenneth Arrow, "Discounting, Morality, and Gaming," in *Discounting and Intergenerational Equity*, ed. P. R. Portney and J. P. Weyant (New York: Resources for the Future, 1999), 13–21, at 14.

30 Posner and Weisbach, *Climate Change Justice*, 149.

31 Conversely, if providing some good to everyone will no longer be possible, then institutions will no longer be required to do so: ought implies can, after all.

32 In my discussion, I have not mentioned any of the awful things that governments in the United States, India, China, and elsewhere have done to socially despised people, and to poor women in particular, in the name of controlling population. While this history is of the greatest social importance, it raises no difficult questions in moral theory. There is no difficulty in understanding why practices of forced sterilization and the fever dreams of eugenicists were impermissible. What *is* truly difficult is how to identify and think about the *permissible* means of affecting how many people will be alive at any given time, not merely in general but in the context of the climate crisis. My contention is that for this purpose it will be essential to have correctly identified the central imperative of intergenerational justice, but it will still not be sufficient.

33 For an arresting discussion, see Mark Lynas, *Our Final Warning: Six Degrees of Climate Emergency* (London: 4th Estate, 2020).

34 For a penetrating discussion of some of the issues, see Clare Palmer, "Does Nature Matter? The Place of the Nonhuman in the Ethics of Climate Change," in *The Ethics of Global Climate Change*, ed. Denis Arnold (Cambridge: Cambridge University Press, 2011), 272–291.

35 For a discussion of the utilitarian assumptions that can produce such recommendations, see Jeff Sebo, "Animals and Climate Change," in *Climate Change and Philosophy*, ed. Budolfson, McPherson, and Plunkett, 42–66, particularly 55–60. But see also the opposite suggestion that if total well-being is what matters, "then we seem to be committed to the idea that a world with, say, 10 quintillion 'happy insects' (each of whom, we can stipulate, has a life containing one unit of well-being) is better than a world with, say, 10 billion happy, flourishing humans (each of whom, we can stipulate, has a life containing one million units of well-being). Why? Because the insect world would contain 10 quintillion (1e+19) units of

well-being overall, whereas the human world would contain only 10 quadrillion (1e+16) units of well-being overall" (59).

36 Kevin Kuruc and Jonathan McFadden, "Monetizing the Externalities of Animal Agriculture: Insights from an Inclusive Welfare Function," *Social Choice and Welfare* (2023): 1–24: "Animal agriculture encompasses global markets with large externalities from animal welfare and greenhouse gas emissions. We formally study these social costs by embedding an animal inclusive social welfare function into a climate-economy model that includes an agricultural sector. The total external costs are found to be large under the baseline parameterization. These results are driven by animal welfare costs, which themselves are due to an assumption that animal lives are worse than nonexistence. Though untestable—and perhaps controversial—we find support for this qualitative assumption and demonstrate that our results are robust to a wide range of its quantitative interpretations. Surprisingly, the environmental costs play a comparatively small role, even in sensitivity analyses that depart substantially from our baseline case. For the model to find that beef, a climate-intensive product, has a larger total externality than poultry, an animal-intensive product, we must simultaneously reduce the animal welfare externality to 1% of its baseline level and increase climate damages roughly 35-fold. Correspondingly, the model implies both that the animal agriculture sector is much larger than its optimal level and that considerations across products ought to be dominated by animal welfare, rather than climate, effects."

37 For an illuminating discussion of this problem, see Mathias Frisch, "Modeling Climate Policies: The Social Cost of Carbon and Uncertainties in Climate Predictions," in *Climate Modeling: Philosophical and Conceptual Issues*, ed. Elisabeth Lloyd and Eric Winsberg (London: Palgrave Macmillan, 2018), 413–448.

38 Sudarshan Varadhan and Aaron Sheldrick, "COP 26 Aims to Banish Coal. Asia Is Building Hundreds of Power Plants to Burn It," *Reuters*, October 31, 2021, www.reuters.com.

39 Roger Harrabin, "UK Seeks to Drill More Oil and Gas from North Sea," *BBC News*, March 24, 2021, www.bbc.com.

40 James Osborne, "U.S. Oil Production Passes 12 Million Barrels a Day," *Houston Chronicle*, July 8, 2019, www.houstonchronicle.com.

41 "Germany to Build Its Own LNG Terminals at 'Tesla Speed' in Shift Away from Russian Gas," *Climate Home News*, April, 28 2022, www.climatechangenews.com. See also Robert W. Howarth, "The Greenhouse Gas Footprint of Liquefied Natural Gas (LNG) exported from the United States," *Energy Science & Engineering* (2024): 1–17, at 1: "Overall, the greenhouse gas footprint for LNG as a fuel source is 33% greater than that for coal when analyzed using GWP 20 (160 g CO_2-equivalent/MJ vs.

120 g CO_2-equivalent/MJ). Even considered on the time frame of 100 years after emission (GWP 100), which severely understates the climatic damage of methane, the LNG footprint equals or exceeds that of coal."

42 Victor Tachev, "Oil and Gas—Current State and What Lies Ahead for the Industry," *Energy Tracker Asia*, June 22, 2022, www.energytracker.asia.

43 George Tubei, "Uganda Is Planning to Sink as Much as $20 Billion in Drilling Over 500 Wells to Expand Its Nascent Oil Industry," *Pulse Live*, January 24, 2020, www.pulselive.co.ke.

44 Charne Hollands, "Top International Oil Companies Exploring Namibia's Hydrocarbons," *Energy, Capital & Power*, February 10, 2022, www.energycapitalpower.com.

45 "The Gas Discoveries Off South Africa's Coast Could Be 'Game Changers'," *Business Tech*, February 11, 2021, www.businesstech.co.za.

46 Lorne Stockman, *Petroleum Coke: The Coal Hiding in the Tar Sands* (Washington, DC: Oil Change International, 2013).

47 Adam Scott and Greg Muttitt, *Climate on the Line: Why New Tar Sands Pipelines Are Incompatible with the Paris Goals* (Washington, DC: Oil Change International, 2017).

48 Jesse D. Gourevitch, Carolyn Kousky, Yanjun (Penny) Liao et al., "Unpriced Climate Risk and the Potential Consequences of Overvaluation in US Housing Markets," *Nature Climate Change* 13 (2023): 250–257.

49 For discussion, see Catriona Mackinnon, "The Panglossian Politics of the 'Geoclique'," *Critical Review of International Social and Political Philosophy* 23 (2020): 584–599.

50 Lamar Smith, "Geoengineering: Innovation, Research, and Technology," Statement by the Chairman of the House Committee on Science, Space, and Technology, November 8, 2017, https://science.house.gov.

51 Conventional climate modeling assumes that by the year 2050, ten billion tons of carbon dioxide will be successfully removed from the atmosphere annually, using an as-yet entirely non-existent fleet of energy-hungry direct air capture facilities numbering in the tens of thousands to tens of millions, and consuming an amount of electricity (50 exajoules) that is equivalent to one-half of humanity's present total annual energy consumption from all sources. See David Kramer, "Negative Carbon Dioxide Emissions," *Physics Today* 73, no. 1 (2020): 44–51, at 46, 49: "The February 2019 National Academies of Sciences, Engineering, and Medicine committee report *Negative Emissions Technologies and Reliable Sequestration: A Research Agenda* concluded that achieving Paris goals without retarding economic growth will likely require that 10 billion tons of CO_2 be extracted from the atmosphere annually by 2050, and that figure will need to increase to 20 billion tons annually by 2100. . . . Large-scale deployment of DAC [Direct Air Capture technology], however, will require enormous

amounts of energy. One study published in July found DAC could constitute as much as a quarter of the world's total energy demand by 2100."
See also the mentioned study, Giulia Realmonte, Laurent Drouet, Ajay Gambhir et al., "An Inter-Model Assessment of the Role of Direct Air Capture in Deep Mitigation Pathways," *Nature Communications* 10 (2019): 6–7: "If large-scale plants are going to be built, capturing 30 $GtCO_2$/year means installing 30,000 facilities. This is comparable with the cumulatively produced number of jet aircraft (21,000 in 1958–2007) or natural gas plants (15,000 in 1903–2000) built in the past. By contrast, considering Climeworks's design, around 30 million units could be required in operational stock by the end of the century. This is aligned to the world annual market for cars and commercial vehicles (73 million units in 2017). . . . DACCS will have a significant impact on global energy provision. In 2100 it could require around 50 EJ/year of electricity, that is, more than half of today's total production (and about 10–15% of the global generation projected in 2100 by our models) and 250 EJ/year of heat, representing more than half of today's final energy consumption globally."

52 "The chief executive of Occidental Petroleum, one of the largest U.S. oil companies, touted its $1 billion-plus project in West Texas to remove carbon dioxide directly from the air . . . 'We believe that our direct capture technology is going to be the technology that helps to preserve our industry over time,' Hollub told the audience. 'This gives our industry a license to continue to operate for the 60, 70, 80 years that I think it's going to be very much needed.'" See Ben Lefebvre, "Oil Industry Sees a Vibe Shift on Climate Tech," *Politico*, March 8, 2023, www.politico.com.

53 For a way of conceptualizing some of these questions, see Britta Clark, "How to Argue About Solar Geoengineering," *Critical Review of International Social and Political Philosophy* 40 (2023): 505–520.

7

NONCONSEQUENTIALISM AND CLIMATE CHANGE

F. M. KAMM

Making People Better or Worse Off

In his excellent chapter for this volume, "On the Moral Challenge of the Climate Crisis," Lucas Stanczyk aims to develop what he calls a "nonconsequentialist, person-regarding" ethic for climate change. Briefly stated, it is nonconsequentialist in focusing on the rights of persons rather than maximizing utility overall; it is person-regarding in focusing on how policies make individuals better or worse off than those very individuals would otherwise have been. Its practical recommendations are to restrict "luxury consumption" and control population growth. In this chapter, I will ultimately consider to what degree Stanczyk's "nonconsequentialist" ethic is nonconsequentialist and also whether it succeeds in its person-regarding aims in different scenarios. I will also propose an alternative nonconsequentialist view. But first, I will review how he arrives at his ethic.

Stanczyk rejects a utilitarian consequentialist approach at least when it tells us to maximize total expected utility. This is because: (1) it can imply maximizing total utility by greatly increasing the size of the human population even though each person will have a life only barely worth living (thus yielding the so-called Repugnant Conclusion); and (2) it can imply imposing enormous sacrifices on the current generation because investing what it would consume will yield greater overall utility by increasing utility in the future. He denies that one generation has duties to maximize overall utility by (a) accepting enormous burdens in order to satisfy all preferences of future generations and (b) creating more people to increase total utility. He introduces rights to show that, contrary to utilitarianism, we do not have a duty to do whatever is best for future generations, only to see to it that they live in a world where their rights are respected.

Stanczyk also rejects appealing to the current self-interest of countries to do what will prevent them from suffering (what I will call) the "direct" bad effects of climate change (e.g., fires, floods, poor air quality, etc.). In arguing against this approach, he says it is not "in the narrow self-interest of the vulnerable countries to take the steps that would be necessary to stave off the worst effects of climate change. That's because the worst effects of climate change . . . will take many centuries to unfold, long after everyone alive today is dead."

He also seeks to make dealing with the Non-Identity Problem "irrelevant" to an appropriate ethic for climate change, though he does not deny facts that underlie the Problem. These facts are that future people will not be harmed (that is, they will not be worse off than they would otherwise have been) by the negative effects of climate change. This is because if we hadn't followed policies that lead to the bad effects of climate change, those people probably wouldn't have existed at all, so they can't be worse off than they otherwise would have been. In other words, instituting policies that prevent the bad effects would result in numerous different interactions and behaviors, different matings and births. In turn this would result in certain future people who would otherwise have existed to not exist and others coming into existence instead. It is those different people who would have lives without the bad effects of climate change though they would not be better off than they would otherwise have been since if the bad effects had occurred they wouldn't have existed. (Also, as long as those whose very existence is conditional on policies that lead to the bad effects of climate change have lives worth living, it seems hard to see how they would have been wronged in being created given that they couldn't have been made better off.)

In addition, Stanczyk argues that just institutions need not be undermined by the bad effects of climate change. However, conditions will be more difficult (e.g., less food to distribute), making it hard to conduct themselves in a just manner.

Given these facts, Stanczyk says:

> From the perspective of a nonconsequentialist and person-regarding view of intergenerational justice, it's not clear why the present generation has a duty to rapidly reduce its consumption growth even if this is necessary to avoid triggering climate tipping points that will eventually prove catastrophic. For it appears that the people who will live through such catastrophes will not be benefited if we do, whereas the duty not to undermine the possibility of just institutions does not seem to rule out continuing with business as usual either. So from a non-utilitarian and person-regarding point of view, how should we reason about intergenerational justice?[1]

In seeking an appropriate "person-regarding" climate ethic, Stanczyk is trying to find an alternative to the argument that even if we couldn't affect particular future persons for the better by preventing climate change, we should create better-off rather than worse-off people. This is known as a "non-person-affecting approach" since the identities of the people who are better and worse off change and no person is affected for the better or worse. (Sometimes this is described as a wide—rather than narrow—person-affecting view since it does not merely interpersonally aggregate all benefits but calculates benefits within individual lives.) Stanczyk's "person-regarding" ethic seems to be what is more commonly known as a "person-affecting" ethic in that he seeks to show how preventing or not preventing climate change can make particular people better or worse off (i.e., holding the identity of the people constant and affecting them for better or worse).[2]

Stanczyk emphasizes person-affectingness and individual rights in his account of nonconsequentialism. However, it should be noted that nonconsequentialism does not deny that consequences, including non-person-affecting consequences, could matter morally. It just denies that consequences are all that matter morally.

How Should Generations Account for Their Part in the Impact of Climate Change?

I will now consider (A) what agents might be responsible for causing climate change, (B) whether the person-regarding effects on which Stanczyk relies in his alternative argument bear on his rejection of the argument for controls on climate change based on the self-interest of vulnerable countries, (C) the different types of person-regarding effects that play a role in his argument and their relative seriousness.

A. One part of Stanczyk's argument focuses on the actions and policies of what he calls the "power elite" or "economic elite" who indulge in luxury consumption of extremely carbon emissions–intensive goods and services (for example, meat eating and air travel). One concern I have about his attribution of "luxury" consumption to the power elite is that the problematic activities he cites are indulged in by the vast majority of people, at least in developed countries. Indeed, it is often members of the middle and even lower economic classes (who are not part of the power elite) who are avid meat eaters and support policies that promote and require use of fossil fuels. If this is so, then members of these populations are not merely the victims of some elite class. Rather, they may be reciprocally harming each other (e.g., in using cars and aviation they are each increasing the emissions to which they are each exposed). Hence, Stanczyk might have to expand the groups responsible for climate change beyond the power elite.

Further, it is said that eating red meat causes many illnesses that can shorten one's lifespan. If this is so, then people who are carnivores may be doing more harm to themselves than would come to them from the effects of climate change in their lifetime. If they are willing to exchange length of life for quality of life (e.g., continue to enjoy consuming meat at the cost of a shorter lifespan) through their own actions, they might also be willing to accept greater risks from climate change caused by others if they are provided with compensating benefits. Even if they have a right against the worst effects of climate change (as Stanczyk holds), they might waive their rights for these other benefits.

B. It is crucial to Stanczyk's person-regarding climate ethic that some of the people whose rights will be negatively affected by actions and policies that perpetuate climate change already exist. He focuses on the children "born yesterday" who in several decades will experience ill effects of climate change. (That is, these are not people whose identities would change depending on what climate policies will be followed.) He says:

> It's false that climate disasters attributable to today's emissions will befall only people who will be born in the future. On the contrary, some children who are already alive will one day be gravely harmed by the emissions that the global economy will put out tomorrow. We have no way of knowing who these particular children are. But even if the lag between emissions and destructive environmental consequences is decades long, we can be certain that some such children are already alive and that they will eventually be gravely harmed.[3]

However, this claim, which is crucial in his argument (so I shall refer to it as the "crucial claim"), seems inconsistent with his view that it is not "in the narrow self-interest of the vulnerable countries to take the steps that would be necessary to stave off the worst effects of climate change. That's because the worst effects of climate change . . . will take many centuries to unfold, long after everyone alive today is dead." If at least some people already alive will be "gravely harmed" in their lifetime, this seems to weaken his grounds for rejecting an appeal to self-interest of the most vulnerable countries to prevent climate change. I will put this concern about an inconsistency in his argument to one side in what follows, but I will assume that the crucial claim is true.

C. Stanczyk thinks that every person once they exist has rights not to be subject to famine, drought, resource wars, and other dire direct effects of climate change. (Suppose for the sake of argument that Stanczyk is correct about what rights people have once they exist.) It is protecting these rights that requires people to undertake burdens to prevent climate change. However, these burdens may themselves involve limiting some rights (e.g., limiting rights to travel and to procreate). He says:

> As a matter of justice, all living persons have claims of varying degrees of moral importance . . . institutions must be erected that continuously give every living person's diverse claims the appropriate weight as soon as and for as long as they live . . . by failing to cut much of the world's present-day luxury consumption, the older generation in charge today in all of the world's countries is wronging some of the children born around the world yesterday. For it will create a situation in which, later in life, not all of these children will be able to escape conflict, famine, and drought, for no reason

more important than the present-day consumption of luxury experiences and goods.[4]

It is important to Stanczyk's person-regarding ethic that there is another way that the children already alive can be affected by future actions and policies of adults that lead to climate change other than the direct bad effects of climate change. His examples illustrate how those children (call them Gen1) could have to bear the burden when they grow up of preventing climate change from negatively affecting the rights of the next generation of children (call them Gen2) who will already be alive during the lifetimes of Gen1 and before some of Gen1's climate-affecting behavior. (I will refer to these burdens of prevention as "indirect" bad effects on Gen1 of climate change. Hence, they will suffer burdens *from* climate change and also suffer burdens *to* prevent climate.) Stanczyk claims that the indirect burden on Gen1 in terms of restricting consumption and reproduction will be greater than the burden to prevent climate change would have been had those who were adults when Gen1 were children reduced luxury consumption. And those who were adults when Gen1 were children may also have been indirectly harmed (in having to bear burdens to prevent climate change) by the previous generation of adults who could even more easily have prevented climate change. (Stanczyk discusses generations going back only fifty years from today, perhaps because he thinks that it was only at that point that it was reasonable to think people should have known the effect of their actions on the climate.)

When Gen1 become adults they will face greater indirect burdens (than the preceding generation would have had to bear) to prevent not only the direct burdens on Gen2 of Gen1's acts but also the additional indirect burdens Gen2 would have to undergo to prevent further rights-violating effects of climate change on Gen3 children who will preexist some of Gen2's climate-affecting behavior. Stanczyk refers to "kicking the can" of increasing burdens to stop climate change down the road as the fundamental problem of intergenerational justice.

Stanczyk's additional view is that "giving appropriate weight to diverse claims" ("claims" being interchangeable with "rights") and deciding which to limit to protect other claims is a matter of considering the seriousness of the interests protected by those claims and how serious the setback of those interests would be if the claim is not protected. For example, he says:

> [I]f our theories tell us that everyone's right to avoid unnecessary famine is more urgent or basic than anyone's claim to eat grass-fed beef or travel long distances by plane for vacation, then we already know what we are morally required to do as a matter of intergenerational justice: Dramatically restrict emissions from animal agriculture, and dramatically restrict aviation emissions from tourism until a truly carbon-neutral alternative to jet fuel is invented. Otherwise, we will be wrongly indulging less important interests in luxury consumption at the expense of survival interests that are comparatively much more important.[5]

Rights Consequentialism

Given all this, one of my questions about Stanczyk's nonconsequentialist person-regarding proposal is how nonconsequentialist it is. Stanczyk's proposal specifically rejects the form of consequentialism that (i) evaluates the goodness of outcomes in terms of whether overall utility is maximized, and (ii) claims there is a duty to maximize overall utility. However, there are other forms of consequentialism that employ a non-utilitarian theory of value with which to evaluate the goodness of outcomes. There are also other characteristics typical of a nonconsequentialist theory besides rejecting that only consequences matter morally. Keeping these additional considerations in mind, I will try to evaluate how nonconsequentialist Stanczyk's view is. I ask this question without assuming nonconsequentialism is correct or aiming to defend it.

A. Is it consistent with nonconsequentialism to decide which claims should be restricted or sacrificed by the strength of the interest they protect? Is this the same as ranking the significance of claims by the interests they protect?[6]

For example, the cases that Stanczyk discusses involve someone or some people doing something, such as flying in planes, that causally contributes to direct bad effects of climate change on others (e.g., deadly fires). This can be classified as doing what (unintentionally but foreseeably) harms those others when it affects for the worse those who already exist at the time of the act. (Though he speaks of rights not to be subject to "avoidable" floods and fires, he never discusses cases in which these events would arise without human intervention such as from meteors landing on Earth but when the bad effects could be prevented by human intervention.) He thinks that restricting a right to travel for pleasure sets back someone's interests less than being killed in a fire and for this reason what he thinks is a right not to be killed in avoidable fires takes precedence over the right to travel by plane.

However, from a nonconsequentialist point of view, when one's actions involve harming others, one might have to sacrifice one's *stronger* interest to prevent others from suffering a setback of their *weaker* interest. An example on the interpersonal level could involve someone who wants to use a machine on their property that will make their business successful and bring them (or prevent their losing) $100,000 income. However, the pollution caused by the machine causes side-effect damage of $50,000 to their neighbor's property. In another version of this case, the side effect of using the machine is that its pollution interferes with the functioning of their neighbor's pump that would have brought that neighbor $50,000 income. In each of these cases, the fact that running the machine would cost the neighbor less than not running it would cost the machine's owner does not on its own imply that the neighbor should bear the cost rather than the person who would use the machine. The moral significance of the separateness of persons on a nonconsequentialist perspective implies this since it implies that bearing a lesser cost for one's own greater benefit is morally different from imposing a lesser cost on another person for one's own greater benefit. Now consider an example involving intergenerational effects. Suppose the current generation would have to give up the right to engage in much future aviation to avoid causing fires due to climate change that will directly affect children already alive when they become adults. Further, suppose (contrary to Stanczyk's surmise about future costs rising) that those children when adults could prevent those fires by giving up a right that would cost them somewhat less than giving up much aviation would cost the earlier adults. On a nonconsequentialist view, we may not be justified in making the potential victim of the harmful act (or third parties who could help them) pay to prevent the fires simply because the cost to them would be less than the cost of not engaging in aviation or preventing the fires would be to those whose behavior would cause the fire threat. Hence, in cases that involve harming, on a nonconsequentialist view it is not necessarily true that the one who should pay to prevent the harm is whoever would sacrifice a claim involving a lesser interest in order to prevent the harm.

Nevertheless, there may be cases in which those who would be victims have previously themselves been harmers and even victimized those who would now harm them. For example, current prominent users (X) of fossil fuel may harm countries (Y) that had previously prominently used fossil fuel that harmed X or other countries. Indeed, the current use of fossil fuel by X may have significant harmful effects only because it builds on Y's past use of fossil fuel. Suppose those whose past acts were harmful to people already in existence at the time they acted should have avoided those acts even at greater cost to themselves than the eventual cost to their victims. It might be consistent with nonconsequentialists' view of harming that those past harmers now pay a greater cost (e.g., in restrictions on claims to engage in activities) to prevent bad effects to themselves of activities of those who currently use fossil fuels, especially if the current users were their past victims. Whether this is so would require further argument (especially to show that it involves person-regarding factors). But it would be different from arguing that those whose claim involves a lesser interest should, simply because of that, have their claim infringed to prevent harm.

B. Nonconsequentialists also typically draw a moral distinction between harming and not aiding, there being a stronger duty to not harm than to aid. So consider a case in which adults in one generation (which I will call the parent generation, or GenP) do nothing to causally contribute to direct effects of climate change. However, the children of their generation would suffer direct bad effects of climate change when they become adults due to the acts of the grandparent (or GenGrand) generation, effects that hadn't yet come to fruition when GenP were children. Suppose the cost (in terms of limitations on rights) to GenP that did not causally contribute to climate change to prevent these direct bad effects of climate change would be less than the cost to the children's generation. GenGrand is not available to bear the cost. From a nonconsequentialist point of view, GenP would not have a duty to pay the cost to help the children's generation merely because it was less than the cost the children's generation would have to pay to help themselves. The nonconsequentialist view is that you need not pay a price (by yielding a right) to help others simply because others would suffer infringement of a more serious right if you did not help than you would suffer if you did help. For example, if you didn't cause a threat to someone's life, you are not obligated to give up your arm (and others may not take it) to save that person's life even from a rights-violating aggressor, though you have the option of making the morally admirable sacrifice. Similarly, you are not obligated to give up your arm to save that victim rather than (i) their giving up an arm and a leg to save their own life or (ii) a third party volunteering to give up an arm and a leg to save them. This is supposed to be consistent with it being reasonable for a rescuer to save someone's life (or their arm and leg) rather than another person's arm when both people cannot be helped because the interest at stake in the loss of life (or arm and leg) is greater than the interest at stake in the loss of an arm. However, ranking the seriousness of rights by the interests they involve does not determine that this rescuer has a duty to sacrifice (or that others may infringe) their less serious right to prevent someone else's loss of a more serious right in cases that involve merely aiding (rather than not harming) others.

Indeed, Stanczyk's view that the ranking of rights by interests does determine the duty to forego exercising a right (or permission to infringe it) might be described as a form of "rights consequentialism." This is a form of consequentialism that involves treating the protection of rights in accordance with their seriousness as the good to be maximized. It is a consequentialist theory of deciding what it is right to do combined with a non-utilitarian theory of value with which to evaluate the goodness of consequences. It thus allows one to sacrifice someone's less serious right as a means of protecting someone else's more serious right. Hence, that Stanczyk's proposal emphasizes rights does not necessarily imply that the proposal is nonconsequentialist. By contrast on a nonconsequentialist view, rights are treated more like side constraints on maximizing the good. Hence, respecting less serious rights of some people might stand in the way of maximizing the good of protecting more serious rights of other people.

Stanczyk speaks of more and less serious claims based on the interests they involve. However, he usually contrasts what he thinks are claims that involve strong interests in absolute terms (e.g., not to die in fires) with claims that involve trivial interests in absolute terms (e.g., eating meat). Sacrificing claims involving trivial interests could be required by nonconsequentialism even in cases of merely aiding. But this would not imply that all claims involving lesser interests should be sacrificed to avoid nonfulfillment of claims involving stronger interests. Hence, especially in cases of simply not aiding, it would be better to attend to the significance of sacrifices in absolute terms rather than merely in relative (less or more serious) terms.

Furthermore, while Stanczyk speaks in terms of infringement of less serious or more serious rights, one should keep in mind that not all costs to protect serious rights will themselves involve infringement of rights. For example, having a high six-figure income (in US dollars) is not a right. Suppose it would cost a later generation 25% of that income to protect the climate-related rights of people already alive but it would cost an earlier generation 10% of that income to protect those rights of the same group of people. On a nonconsequentialist view, it is not the duty of the earlier generation to pay the lesser cost instead of the later generation having to pay the greater cost to prevent the rights violation merely because of the difference in cost. Sometimes it can be unclear whether a case involves harming or not aiding. For example, Stanczyk discusses a case in which for the sake of "indulging runaway luxury consumption," grandparents overfarm their land. This leads to their children (GenP) (who existed before the consumption occurred) being unable to adequately feed more than the two children they already have (Gen1), though they would like to have a third child. Stanczyk thinks that the parents (GenP) would wrong each other and their already existing two children in having the third child if it led to severe rationing for the family (thus harming them relative to conditions without the third child), but allowing the new child to starve would wrong that child. In Stanczyk's view, limiting their reproductive rights is a greater burden on GenP than limiting a liberty right to consume luxury goods would have been for the grandparents, but restricting reproductive freedom involves sacrificing a lesser interest than the one protected by the right to adequate nutrition.[7] With respect to the situations created by both the grandparents' indulgence and GenP if they have the third child, Stanczyk says: "In creating this situation, moreover, they would be acting without adequate justification, because they would be privileging less important interests over interests that in the circumstances were even more important."

One puzzling feature of this case is Stanczyk's claim that if the parents had their third child, they would be wronging each other. It is not clear how this could be so if they agree to have another child in full knowledge that they would have to reduce their own portion of food in order to feed another mouth. This is because they could be waiving their right to an equal share of nutrition in order to have another child. Of course, they may still be harming their already existing children by reducing those children's food in a "painful system of food rationing" in order to avoid letting the new child starve (and thus fall below a life worth living). Suppose the third child could have a life worth living without getting an equal share of food and would not be made worse off than it would otherwise have been if its creation were conditional on its getting less than others. Suppose further that the third child receiving less than others would eliminate the need to institute a painful rationing system on the existing children (or even the parents). Then it is only if all the persons in the case had a right to an *equal* share of food that anyone would be wronged.

Could we interpret Stanczyk's case as involving GenGrand simply not aiding (rather than harming) GenP? A clear case of not aiding could involve GenGrand "indulging" in leisure while living frugally rather than farming their land to its true potential, as a consequence of which (assume) it will be much harder for their children (GenP) to make their land productive enough to support more than two children consistent with feeding everyone adequately. Though it would have been less of an imposition on GenGrand's interests for them not to have so much leisure than for GenP to limit their reproduction, on a nonconsequentialist view this need not imply that the grandparents had an obligation to help GenP by working harder. For another example of clear not aiding, suppose the leisure is required to save GenGrand's arm from severe damage in their working harder. Saving one arm is a lesser interest than saving an arm and leg. Yet the grandparent would have a right not to do farm work that would cause the loss of their arm even if their working would in itself prevent the land not producing enough to keep someone in GenP already in existence from losing an arm and leg.

By contrast to these cases, could the grandparents in Stanczyk's actual case be understood as simply not aiding because their lavish consumption diminishes the inheritance they leave for GenP? This might make them be only like parents who permissibly use up all their money, leaving little inheritance for their children who have no way of becoming as well-off as their parents. However, in Stanczyk's case the grandparents overfarm and ruin the land in order to attain their wealth. If one thinks of the land as not belonging to GenGrand but to the community or family over generations, then GenGrand's behavior may involve wrongfully harming because they negatively affect the land. By doing that, they negatively affect the lives of others who co-own the land and have a right to prevent its being ruined to their detriment.

Stanczyk's case involves particular parents and their children (rather than general parental and offspring generations in a society). In this context obligations between different generations are probably stronger than between nonfamily members of different generations. This factor should be considered before drawing conclusions about general intergenerational justice from a family case. On the other hand, it might be argued that even if family members over different generations are not co-owners of land, society is entitled to grant personal ownership only on condition that the land is farmed in a responsible way. Such an argument might be necessary to account for a duty not to overfarm for one's own benefit when this leaves less usable land for the next generation. Indeed, Stanczyk's case may be taken to involve violation of an intergenerational Lockean proviso in which the grandparents take for their sole use more of the world's resources than is consistent with "enough and as good" being left for others.[8]

C. What has been said about aiding from a nonconsequentialist point of view may also bear on Stanczyk's claim that he provides a person-regarding justification for duties to prevent climate change for each generation over the entire course of history. He says:

> The reasoning illustrated is able to establish an unending chain of person-regarding non-utilitarian intergenerational obligations that is indifferent to the identity of future generations. It is sometimes said that thinking about intergenerational ethics exclusively in terms of what we owe to our younger contemporaries ignores the independent claims of as-yet unborn future generations. But as I have just shown, that is a mistake. It is an essential part of the explanation for wrongdoing that I have given that future people *will* have moral claims on everyone as soon as they come into existence and that—unless decisive steps are taken in the present—even greater sacrifices will then be morally necessary to honor these rights, on pain of grave injustice.[9]

Is this unending chain of person-regarding obligations present in the following case? Suppose that some people (GenP) in the presence of already existing children engage in activities that contribute to climate change. However, the direct negative effects are delayed so that neither these children (Gen1) nor the next generation of children (Gen2) are affected. Further assume that when Gen1 and Gen2 become adults they will not add to climate change. They do not undergo great sacrifices to make this happen but only have a happy lifestyle that either does not contribute to climate change or requires only minimally costly alterations to prevent contributions. In this case, GenP's behavior does not directly negatively affect already existing people (Gen1) or Gen2 people but only Gen3 people. People in Gen2 and Gen3 would not already have existed when GenP acted and they may owe their worthwhile existence to GenP's acts that result in eventual effects of climate change. Hence they will not be worse off as a result of GenP's acts than they would have been had GenP acted differently since they wouldn't then have existed. As noted, one part of Stanczyk's argument emphasizes that those already in existence at the time of the climate-changing behavior will be directly harmed by that behavior because the direct effects of climate change will eventually occur (perhaps decades later) in their lifetimes. This crucial claim will not hold in the case I have described.

However, is it possible that those already existing when GenP acts, whose identities would not change based on what GenP does (i.e., Gen1), suffer indirect person-regarding negative effects from GenP's acts? This would be so if Gen1 would have to pay to prevent rights violations to Gen3 that would occur when they exist (if they have a right not to be exposed to fires, floods, drought, etc.). Presumably the argument for this might be as follows: For Gen2 to prevent conditions that will violate Gen3's rights will be very costly in terms of restrictions on its rights. It would be less costly for Gen1 to prevent the violations. Hence, instead of kicking the cost-can down the road to Gen2, Gen1 should bear the cost. But that means that Gen1, who already existed when GenP acted in ways that resulted in climate change, will be made worse off as a result of GenP's actions. They may not suffer *from* climate change but they would suffer *to prevent* its effects on Gen3.

However, this argument depends on (i) Gen1 and Gen2, who did nothing to contribute to climate change, having to bear burdens so that Gen3's rights are respected, and (ii) Gen1 having to bear burdens for Gen3's sake simply because those will be less than Gen2's and Gen3's burdens would be. It has already been suggested that nonconsequentialism might not endorse the reasoning involved in (ii) and that it might not endorse (i) since it distinguishes responsibilities based on avoiding (or compensating for) rights-violating harm one causes from responsibilities to merely aid to prevent such harm others cause. Possibly GenP might have responsibilities to prevent violation of Gen3's rights not to be subject to floods, droughts, etc. (e.g., by investing in a fund for use after GenP passes away). But this obligation would not be based on person-regarding considerations since Gen3 did not already exist when GenP acted, and identities of those in Gen3 may have depended on GenP's having acted as they did. If this is so, GenP's obligations would not be based on its having harmed Gen3 since actual members of Gen3 wouldn't have been better off if GenP hadn't acted given that they wouldn't have existed. Hence, dealing with the Non-Identity Problem would not be made irrelevant to deriving any duties of GenP's, as Stanczyk wishes.

Ordinarily it would seem to be good if we could easily delay the direct bad effects of climate change. This might be in part because new inventions will make it easier rather than more burdensome to deal with such effects. Also, assuming the direct negative effects would not eventually peter out, fewer generations will be negatively affected if there is a delay. Ordinarily, it is also good if intervening generations between original climate-change actors and those directly affected by climate change do not contribute additionally to climate change. However, in cases where costs of preventing bad effects increase as time goes on, the two factors of delay in bad direct effects and innocent intervening generations combined with nonconsequentialist concerns about limits on duties to (merely) aid may create problems for establishing "an unending chain of person-regarding non-utilitarian intergenerational obligations that is indifferent to the identity of future generations."

A variation on this case raises an additional issue. Suppose GenP arranges for the bad direct effects on climate of its acts to be delayed until Gen3 but this makes their effects much worse for Gen3 than the effects would have been if felt by Gen1, or even if felt by both Gen1 and Gen2. Those in Gen3 would not have existed with (assume) lives still worth living had GenP not acted as they did. In this case, Stanczyk's approach implies that to avoid person-regarding harms and wrongs to people in Gen1 already in existence when GenP acted, GenP should do what imposes a much worse burden in absolute terms on people in Gen3. This seems morally problematic even if one should reject what Parfit called the No-Difference View, according to which there is no moral difference between (1) creating a person who is worse off than someone else one could have created and (2) making someone worse off to the same degree when *that person* could have been better off. Contrary to the No-Difference View, it seems morally worse to produce pollution that will cause someone to die at age seventy rather than seventy-five by contrast to producing pollution that will lead to creating someone who will die at age seventy rather than someone else who will die at seventy-five. However, rejecting the No-Difference View does not settle what to do when the person-regarding harm would be much less than the difference between the person who would be created to a worse life and someone else one could have created to a better life. Deciding whether and to what degree to give preference to the interests of Gen1 by person-regarding acts rather than to Gen3 by non-person-regarding acts will require dealing with the Non-Identity Problem.[10]

D. Consider another factor that might bear on whether those who do not contribute to climate change have a duty to aid those who would be affected by climate change caused by others. Even if Gen1 and Gen2 in my previous example do not act in ways that cause climate change and are not directly negatively affected by GenP's acts that cause climate change, they might still be beneficiaries of those acts of GenP's (e.g., by having a higher standard of living enabled by GenP's use of fossil fuels). Suppose one could show that GenP's acts were impermissible because they led to the existence of Gen3 people bearing direct burdens of climate change instead of different people (Gen3b) without such burdens. Then Gen1 and Gen2 might have a duty to forfeit some of the benefits to them of GenP's impermissible acts (those benefits being like ill-gotten gains) and to aid Gen3 by transferring those benefits to them. However, such an account of Gen1 and Gen2 having the duty to aid Gen3 would not make dealing with the Non-Identity Problem irrelevant. This is because it would require a non-person-regarding account of why GenP's acts were impermissible (so that benefits gotten from them were like ill-gotten gains) even though those acts harmed no one, including those in Gen3, given that they would not have existed with lives worth living but for GenP's acts.

A Possible Alternative Rights–Related View

In conclusion, let me return to the earlier quote from Stanczyk, now distinguishing its parts: "(a) It is sometimes said that thinking about intergenerational ethics exclusively in terms of what we owe to our younger contemporaries ignores the independent claims of as-yet unborn future generations. But as I have just shown, that is a mistake. (b) It is an essential part of the explanation for wrongdoing that I have given that future people *will* have moral claims on everyone as soon as they come into existence, and (c) that—unless decisive steps are taken in the present—even greater sacrifices will then be morally necessary to honor these rights, on pain of grave injustice."

I have argued that, contrary to (a), thinking about intergenerational ethics exclusively in terms of what we owe to our younger contemporaries *can* sometimes ignore the claims of as-yet unborn future generations. I have also tried to show that, at least according to nonconsequentialism, the choice presented by (c), between greater and lesser sacrifices, need *not* always morally require that lesser sacrifices be made. However, it is also important to emphasize that *claim (b) is separable from (a) and (c)* and it may be important on its own in a nonconsequentialist approach to dealing with the Non-Identity Problem and climate change. That is, I agree with Stanczyk that rights are important in the nonconsequentialist approach to climate change but I think he does not sufficiently focus on them. He introduces rights to show that, contrary to utilitarianism, we do not have a duty to do whatever is best for future generations. He thinks we do have a duty to see to it that they live in a world where their rights are respected (though, I have argued, he does not distinguish between a duty not to do what violates their rights and a duty to help satisfy their rights). However, Stanczyk then shifts the focus to an argument for the unending chain of person-regarding relations as a way to deal with climate obligations. Perhaps there is a different rights-related view worth developing that will not seek to make the Non-Identity Problem irrelevant but instead help deal with it.

Here briefly is a possible alternative rights–related view. If we create some people who are worse off than other people we could have created, this alone does not show that we have violated anyone's rights. However, if we create people whose lives are worth living and could not have been better, this also may not show that we have not violated these people's rights in at least some way. For example, suppose we know that an act necessary for creating someone with a life worth living is the very act that unavoidably sets a delayed-action bomb that will kill the person when they are twenty. Call this the Delayed-Action Bomb case. It could be a reason not to create the person in this case that the act needed to create them to a worthwhile life would also be the act that violates their right not to be unjustly killed. We should not feel free to commit the act that will kill someone simply because the person we will kill would not otherwise exist with a life overall worth living.[11]

That is, there will still be an infringement of their right not to be killed caused by the act that creates them even if they have no right not to be created to such a life containing the infringement. Because this infringement exists, the fact that a person might have no rational objection to having been created to such a life may not be enough to put to rest a moral concern from the agent's point of view to creating the person. After all, the agent will be responsible for infringing the person's right at age twenty. One possible reason for an agent giving priority to not performing an act prior to the person's existence that will infringe the person's rights once that person is alive is that there is no one who will be deprived of a benefit, or made worse off, or wronged, if they are not created with a life worth living. On this account, the agent giving priority to not infringing the right is an agent-focused reason since it does not focus on an objection from the point of view of the person created who will be killed (what is called a victim- or patient-focused objection). By contrast, when someone is already in existence, there may be no such reason sufficient for an agent to give priority to not doing the act that infringes that person's right. This is so if doing the act that sets the bomb is the only thing that can provide someone who is already in existence with twenty years of overall life worth living without which they would die immediately and thus be much worse off.[12]

However, this leaves it open that in cases like the Delayed-Action Bomb an agent may also act for the sake of a victim-focused rights-related reason. This is so even if we can imagine that a hypothetical victim himself would sacrifice the right to have a life worth living. The agent may be better positioned to uphold respect for a right that is connected to the status of being a person than someone who has the right, placing respect for that right ahead of serving someone else's interests. This is so when giving priority to non-violation of the right does not endanger the potential victim's interests because there is no one whose interests are negatively affected by not being created. Such an account would be victim-focused by focusing on their right, even if it doesn't focus on the victim's overall interests and their possible complaints in being created. This exemplifies what I have elsewhere described as a concern with "rights beyond interests" though it exemplifies it in a different way than I have elsewhere described.[13] In this version, it is not that the right that would be infringed is important independent of its effect on the person's interests (such as a right not to be lied to). It is that the creating agent can afford to give less weight to the person's *overall interests that would make their life worth living* than the weight the person themselves would give to their overall interests once they exist.

This account takes a perspective on creating that is ex ante to the creation rather than imagining what a person who is created would think from a perspective ex post their creation. It is this ex ante perspective that could also lead an agent not to create someone who would certainly be a slave even though that person would still have a life overall worth living. Ex ante the agent may give priority to avoiding the gross attack on human dignity involved in slavery even if once someone exists they would consent to slavery if it were the only way to have a life still worth living, and an agent acting on their behalf might facilitate this.

Of course, in creating people it could be reasonable to believe that during their lifetime some of their rights will be violated even by those who created them. This possibility is often not a sufficient reason for not creating the people who will still have lives worth living. Hence, the alternative view to which I have gestured faces the problem of distinguishing between the expected rights violations (or even bad states of affairs not caused by rights violations) that do and those that do not sufficiently justify not creating.

Nevertheless, in dealing with the Non-Identity Problem in the context of climate change, it could be important to know what, if any, climate-related rights people *will have once they exist*. This is because an agent may sometimes have both victim-focused and agent-focused reasons for not creating someone when the very act that makes it possible for someone to exist is also an act that will violate their rights once they exist, for example by causing the bomb to go off or by producing emissions that poison the atmosphere. This is so even if in creating someone the agent doesn't make anyone worse off than they otherwise would have been and everyone has a life worth living.[14] These rights-related reasons could give an agent reason to do what will result in their creating a different set of people whose rights would not be infringed even if they would not be better off than they would have been had the agent acted differently. These rights-related reasons might also give an agent reason to create these people even if their lives worth living would not be overall as good as different people the agent could create whose rights would be infringed.[15]

Notes

1 Stanczyk, "On the Moral Challenge of the Climate Crisis," in this volume.

2 In my earlier work I have argued that a moral distinction should be drawn within person-affecting acts, not only between person-affecting and non-person-affecting acts. That is, we could affect a person for the worse by altering the egg, sperm, or embryo from which they arise by contrast to doing something to the person once they exist. For example, we could affect someone for the worse by doing something to the egg from which they will arise that reduces their IQ from 140 to 120. It may be as morally permissible to do this as to not give someone genes for a 140 IQ to begin with. However, it could be morally impermissible to lower one's child's IQ from 140 to 120 (for example by giving them a drug) once they become a person. This is because a person has a right not to be given a drug to eliminate a beneficial characteristic they already have. This is so even if there is no right requiring that persons have a 140 IQ. By contrast, an egg or embryo that is not a person does not have a right to retain beneficial properties. It matters morally *how* we affect a person for the worse. However, some properties are such that it would be wrong to deprive a person of having them (e.g., an IQ of at least 100). In these cases, *how* we affect a person so that they do not have such a property may not matter; we would wrong the future person in depriving them of the property even by affecting the egg from which they arise. I first raised these points in F. M. Kamm, *Creation and Abortion* (New York: Oxford University Press, 1992) and again in (among other places) F. M. Kamm, "Moral Status, Person Affectingness, and Parfit's No-Difference View," in *Rethinking Moral Status*, ed. S. Clarke and J. Savulescu (Oxford: Oxford University Press, 2021). Perhaps Stanczyk uses "person-regarding" rather than "person-affecting" because he thinks the former but not the latter requires a person to already be present. This would cohere with his emphasis (as we shall shortly see) on doing what affects people who preexist climate-changing acts. But it is not clear that he would think that affecting an egg that preexists a climate-changing act (when a person will definitely arise from the egg) can be morally different from affecting a preexisting person.

3 Those who are currently adults may only be affected by behavior of past generations, but insofar as they had already been born before some of that behavior, their identities would also not have been changed by that behavior. Hence, they too might be covered by Stanczyk's argument. See Stanczyk, "On the Moral Challenge of the Climate Crisis," in this volume.

4 See Stanczyk, "On the Moral Challenge of the Climate Crisis," in this volume.

5 See Stanczyk, "On the Moral Challenge of the Climate Crisis," in this volume.

6 Stanczyk discusses rights (claims) that protect interests. He may believe in the Interest Theory of rights and accept that all rights protect interests. I have argued against that view and argued that some rights protect the status of persons independent of what is in their interest. See, for example, F. M. Kamm, "Rights," in *The Oxford Handbook of Jurisprudence & Philosophy of Law*, ed. J. Coleman and S. Shapiro (New York: Oxford University Press, 2002), revised as F. M. Kamm, "Rights Beyond Interests," in F. M. Kamm, *Intricate Ethics* (New York: Oxford University Press, 2007). For purposes of this chapter (unless otherwise noted), I will restrict myself to rights that protect interests.

7 As noted earlier, Stanczyk's view is that reducing consumption and population growth are two important ways to deal with climate change. So it is interesting that this case involves a choice between lavish consumption and an additional child in two different generations. He thinks it is less of a burden to give up the consumption than to give up having the third child, though he might actually prefer that both be given up. It is also worth noting that one person within his own life might well decide to limit reproduction in order to be able to engage in lavish consumption. This is an indication that giving up the value gotten from such consumption could be more of a burden than giving up an additional child at least within one person's life.

8 Michael Otsuka suggested that the grandparents violate an intergenerational Lockean proviso. He discusses a case like this in Michael Otsuka, *Libertarianism Without Inequality* (Oxford: Oxford University Press, 2003).

9 Stanczyk, "On the Moral Challenge of the Climate Crisis," in this volume.

10 This variation on my case was suggested by Michael Otsuka. For more on the rejection of the No-Difference View, see Jeff McMahan, "Causing People to Exist and Saving People's Lives," *Journal of Ethics*, no. 1/2 (2013): 5–35.

11 This view takes on the challenge of the earlier claim mentioned in this chapter that "so long as those whose very existence is conditional on policies that lead to the bad effects of climate change have lives worth living, it seems hard to see how they would have been wronged in being created given that they couldn't have been made better off." In my Delayed-Action Bomb case, the violation of a person's right not to be killed by a bomb is meant to be analogous to the violation of any climate rights people have.

12 In F. M. Kamm, "Moral Status and Personal Identity: Clones, Embryos, and Future Generations," *Social Philosophy and Policy* 22, no. 2 (2005):

283–307, I wrote: "If we wish to do right by future generations, it will be important to know what level of environmental quality they are owed and, surprisingly, when the alteration to the environment will occur relative to the existence of the persons affected by such an alteration." However, what I had in mind here is not dependent on there being a Non-Identity Problem. That is, suppose the identity of future people were fixed regardless of what climate policy is followed. It might still be true that before someone exists the environment is "not theirs" over which they have rights (in a possessive sense) but after they exist, the environment is "theirs." Acts committed before that person exists may diminish the environment they inherit when they are born, however, the environment acted on was not "theirs" but an earlier generation's at the time the act was done. By contrast, acts done after that person exists may violate their rights (that others not act on "their" environment) in a way earlier acts do not. However, if (as Stanczyk believes) everyone has rights not to be exposed to disasters due to climate change, then whether these changes come about by acts that affect an environment before or after a determined-to-exist person actually exists may not matter. This is because acting on what is one's own could be prohibited if it causes violation of other's rights not to be exposed to disasters. See also F. M. Kamm, "Genes, Justice, and Obligations to Future People," *Social Philosophy and Policy* 19, no. 2 (2002), and Kamm, *Intricate Ethics*, chapter 7.

13 See Kamm, "Rights Beyond Interests."

14 In creating people an agent might also (1) give priority to not placing them in the way of others who would infringe their rights (when the placement would necessarily accompany the creation to a life overall worth living); (2) give priority to not creating someone with very bad handicaps (e.g., blind and deaf) in a life still worth living (like Helen Keller's) even if these bad conditions do not involve great suffering or come about through rights infringements. I have elsewhere considered problems that can also be concealed by only considering whether a life is overall worth living, including whether one should undergo long-term suffering even for the sake of an overall greater good. See chapter 2 in Kamm, *Almost Over: Aging, Dying, Dead* (New York: Oxford University Press, 2020) and Kamm, "The Rationality of Suicide," in *The Oxford Handbook of the Philosophy of Suicide*, ed. M. Cholbi and P. Stellino (New York: Oxford University Press, forthcoming).

15 I am grateful to Michael Otsuka, Anja Jauernig, and Nir Eyal for their comments on earlier versions of this chapter. I also thank Chiara Cordelli for including my chapter in this volume and for her assistance.

8

KICKING CANS AND TAKING STOCK

STEVE VANDERHEIDEN

For almost as long as humans have recognized that some increasingly common actions and activities were changing the planet's climate, this anthropogenic phenomenon has been recognized as presenting an ethical challenge with intergenerational dimensions. Destabilizing the planet's climate system, with the range of impacts on human and natural systems that are widely expected to accompany this, is now widely recognized as not merely imprudent but also wrongful. Having discovered what we have done and continue to do to the climate system—and thus indirectly to everyone and everything that shares our planetary residence—we owe it to future persons or generations to mitigate these impacts by reducing the relevant actions or activities as well as facilitating adaptation to the planetary warming that we can no longer avoid causing. As signatory parties to the 1992 United Nations Framework Convention on Climate Change acknowledged in that landmark call to international cooperative action, humanity has a collective ethical obligation to "protect the climate system for the benefit of present and future generations of humankind." Furthermore, state parties to the convention are to fulfill this ethical imperative by ethical means, undertaking remedial actions "on the basis of equity and in accordance with their common but differentiated responsibilities and respective capabilities."[1]

Now, more than one human generation after committing to freezing national emissions at 1990 levels pending further development of an equitable international treaty framework and following two landmark international mitigation treaties in the 1997 Kyoto Protocol and 2015 Paris Agreement painstakingly brought into force under the UNFCCC, that core ethical objective remains elusive. As Lucas Stanczyk observes in his contribution to this volume, despite three more decades of scientific research, climate policy development and implementation at scales from local to global, and public attention to its increasing urgency, humanity is still collectively kicking that can down the road. Unwilling or unable to take those mitigation actions that we have acknowledged as necessary since 1992, most of the state parties to that convention continue to shirk their collective obligation to protect the climate system by failing to commit to sufficient mitigation actions for the mitigation goals that they have endorsed and even largely failing also to develop and implement sufficient mitigation policies and programs to meet their inadequate commitments. In doing so, the current generation imposes on future persons (and/or current persons in the future) some combination of increasingly serious climate impacts that are now certain to cause harm to many no matter what we do from this point forward along with an increasingly difficult and costly challenge in preventing climate change beyond the threshold that UNFCCC processes have identified as compatible with the maximum feasible ambition of the world's existing governments (a level of ambition that has yet to obtain).

Given this, it would seem entirely unproblematic to many nonphilosophers to conclude that by kicking this can down the road, the current generation is failing in its obligations to future persons or generations. Indeed, some of the most important advocates for stronger climate action often express precisely this sentiment.[2] Philosophers, however, have questions. Is there really a collective moral agent that encompasses an entire generation of human persons that can be assigned moral obligations and thus be held responsible for failing to meet them? If there is, who is included and who is not, by what process (if any) is that agency directed, and how (if it does) can the collective responsibility of the entire generation be devolved into responsibilities of some of its constituent collectives (like nation-states) or perhaps even to individual persons? What about the subjects of this obligation, to whom our climate-related duties are owed? Should it be viewed as a single, undifferentiated collective (as in duties to "futurity"), as a set of multiple and differentiated collectives (whether defined in terms of generational membership or otherwise) to which we may have differentiated duties, or to individual persons with specific identities (which then invites objections from the Non-Identity Problem,[3] to which scholars of intergeneration ethics or justice have by now devoted considerable attention).

Setting aside those questions for a moment, what is it specifically that is owed by one agent or set of agents to another moral subject or set of subjects? Is it to refrain from harming them (through "dangerous anthropogenic interference" in the climate system) at all? If this is our objective, then we have already failed. Environmental factors already take the lives of 13 million persons each year, with 23 million displaced annually by weather-related events, much of this resulting from climate change.[4] Likewise, if it includes maintenance of some capacity to avoid harming in the future, as in Stanczyk's application of a principle of disjunctive wrongs to climate ethics in this volume, where "it wrongs people to act in ways that will necessitate either wronging or harming them in the future." If *ought* implies *can*, we cannot now owe future persons (or our younger contemporaries) a world free of climate-related loss and damage, given that such harm is already occurring, and more is now unavoidable. Is it rather some combination of (or priority scheme among) efforts to minimize dangerous climate impacts through reduction in the drivers of climate change (or *mitigation*), efforts to try to shield would-be climate victims from harm that results from the ongoing and expected impacts of a warming planet that we fail to avert (*adaptation*), and then programs to compensate those for whom our mitigation and adaptation efforts fail for the anthropogenic loss and damage that they suffer? If this threefold imperative jointly constitutes that obligation, as I think it does, then many climate victims will have already been failed by mitigation and adaptation failures that have also gone uncompensated or are impossible to correct through compensation, so perhaps all three of those cans are still being kicked down the road. Specifying objectives that cannot be met is unhelpful and perhaps counterproductive—contributing toward climate despair or an apocalyptic fatalism that undermines our motivational capacities to act—but imperatives other than harm avoidance (or nonmaleficence joined with corrective justice for those that suffer loss and damage) require further elaboration and justification.

What follows from this observation about kicking cans down roads? Insofar as climate ethics aims to contribute to public philosophy, its diagnoses of moral failure should do more than reinforce narratives of decline or generate unconstructive outrage at offenders. Ideally, these diagnoses should also be able to chart a path forward, or at least to articulate the various dimensions of the ethical challenge in a manner that contributes to our understanding of it. In order to try to get normative traction on this problem, philosophers ask more critical questions about what follows from the wrongness of can-kicking. Whose responsibility is it to act adequately on climate change, in order to avoid kicking the can down the road, and what specifically must they do? What (if anything) should we in our capacities as individual persons do when we belong to communities like nation-states over which we have relatively little control when these states continue can-kicking at the collective level? How can ethically egregious can-kicking be distinguished from other actions or inactions that it resembles, such that we or others can identify when we have met this obligation to cease this practice and when we have not? How much must we do before we have done enough mitigation, what ought we as individual persons do about actions that can only be done on a collective scale, and how do both of these questions change over time, based on what we will have done in the past and what we may come to know in the future? How can theory inform practice in this context?

In the spirit of public philosophy and its desire to wield philosophical analysis in the service of better understanding and more constructive responses to public problems, I shall begin by contrasting two distinct ways of unpacking the observation about generational can-kicking, arguing for an alternative that is based in mitigation burden-sharing rather than harm avoidance, differentiates responsibilities among collectives and can potentially guide individual as well as collective mitigation actions, and relies on an undertheorized concept (of stock-taking) in climate ethics that may be able to accommodate some dynamic elements of intertemporal obligations that static burden-sharing (which conceptualize demands of climate ethics in terms of a fair allocation of costs associated with protecting the climate system) or harm-based accounts (which conceptualize it in terms of acts that cause or avoid harm) cannot. Finally, I shall attempt to distinguish ethically dubious manifestations of can-kicking from ostensibly similar actions and practices undertaken at individual and collective levels that do not raise the same moral problems.

Can-Kicking, Harm, and Fairness

Stanczyk's account of what might be termed the ethics of can-kicking could be unpacked in at least two different forms, each with its own foundational principle and argument structure. One is through a duty of beneficence or harm avoidance—that we (individually or collectively) have a duty to rescue the vulnerable from climate peril to which we would otherwise subject them as well as one to ensure that persons in the future are able to do the same. Both take the injunction against avoidable harm as fundamental, whether in terms of the direct climate-related harm that our shirking of mitigation causes or the indirect ethical peril of that same shirking making it very difficult or impossible to avoid causing such harm in the future. While the former has been widely treated within the climate ethics literature, Stanczyk captures the latter in his principle of disjunctive wrongs, through which he maintains that "it wrongs people to act in ways that will necessitate either wronging or harming them in the future."[5] From this principle he derives a duty to maintain a future capacity to avoid harm, which we now owe to our younger contemporaries (thereby avoiding the moral thicket of obligations to merely possible future persons including the Non-Identity Problem) but undermine by our can-kicking. He identifies the "core imperative of intergenerational justice" as the duty to "put in place all those restrictions on existing rights and freedoms that, unless they are put in place, will require even more serious sacrifices having to be made later by today's younger people, lest they fall afoul of the very same imperative in the future."[6] Failing to restrict our collective greenhouse emissions, which Stanczyk understands in terms of a failure to restrict certain rights and freedoms, wrongs some persons who are alive today because of the greater sacrifice (presumably of those same rights and freedoms) that they will need to endure in order to avoid generating a given level of climate-related loss and damage. The intergenerational obligation is therefore conceptualized in terms of this indirect (or disjunctive) harming, rather than direct harming (presumed to be impossible given the Non-Identity Problem) or any kind of intergenerational distributive duties. As Aaron Finley casts what he terms "burden dumping," this shirking "constitutes a double wrong—they wrong those they fail to rescue, and they wrong those on whom their burdens fall."[7] Insofar as we foreclose morally decent futures that remain currently possible but depend on our taking adequate mitigation actions in the near future, we commit another kind intergenerational harm in which future persons could become both perpetrators and victims of harm as a result of the same set of current actions.

Such a formulation invites several objections. First, it remains unclear how future persons could be held responsible for harm that is impossible (or prohibitively difficult) for them to avoid triggering due to our actions now. If the relevant existing agents delay or otherwise avoid acting with sufficient urgency and the window to prevent future harm actually closes as a result, then it will have been those current agents that wronged future climate victims. No principle of disjunctive harm would be needed to identify this offense. Likewise, if current can-kicking makes it prohibitively costly for future agents to prevent such harm, at least above the kind of difficulty threshold that Stanczyk's formulation suggests, responsibility for this would again fall on those whose acts and omissions brought this about. On the other hand, if this future cohort in question could still feasibly be expected to meet some kind of mitigation target, even if the costs to them of doing so could have been lower if earlier generations had acted differently and so made this task easier, then it remains unclear how that later cohort could have been wronged. They may think poorly of us for our lack of initiative, and rightfully so, but that is different from being harmed unless each generation is entitled to expect a duty of assistance on a potentially wide range of matters to have been fulfilled by their predecessors and the gap between expected and received assistance was sufficiently wide and disruptive.

Second, such a principle (that it is wrong to make it impossible or very difficult for other agents to avoid doing wrong in the future) implicitly assumes that this window is still open—that if only we do enough now and in the near future then we can still protect others from wrongful direct harm (i.e., from climate change rather than from capacities to mitigate it). Again, morally and quantitatively significant direct harm from climate impacts has already occurred and more will transpire regardless of how much mitigation we undertake now, so the window for preventing serious anthropogenic harm altogether cannot realistically be viewed as still open. Surely it is not too late to *reduce* the frequency, magnitude, and probability of wrongful climate-related harms through more ambitious mitigation undertaken in the near future, while perhaps also easing somewhat the mitigation burdens of future cohorts in pursuit of the same mitigation target, and both present themselves as compelling imperatives of intergenerational justice. But seeking to reduce rather than avoid anthropogenic harm calls for a different ethical principle than is available by defining the goal as harm avoidance, which can no longer serve as a feasible goal.

Economists sometimes identify such targets through calculations of optimal pollution, where the marginal cost of pollution abatement intersects with the marginal benefit of averted pollution-related harm, but efficiency would be a fraught basis for an ethical principle. Such targets are sometimes constructed politically, as with the Paris Agreement targets of capping global temperature rises at "well below 2 degrees Celsius" while also aiming to prevent warming of more than 1.5 degrees, reflecting compromise between those urging stronger protection and those appealing to issues of cost or political feasibility. In this sense they may also lack the robustness of ethical principle but at least stipulate a target that is defensible by reference to its democratic legitimacy. Neither temperature target promises to protect everyone from direct climate-related harm, so neither offers a principled basis for distinguishing between moral success and moral failure. The stricter target represents a more ambitious aspiration that calls for stronger action and promises less serious corresponding harm (although still significant harm) while the latter identifies a politically constructed moral baseline through which ongoing actions can be planned and evaluated, providing a more practically useful form of normative reasoning that is available through prohibitions on direct or indirect harm that is now impossible to avoid.

Another (and, as I shall argue, more promising) formulation of the ethics of can-kicking is also possible, and indeed is suggested by some of what Stanczyk aptly condemns about the current generation's tendency to kick this particular can down the road. Rather than a no-harm principle within an account of intergenerational ethics that is based around duties to identifiable others, this other formulation turns on the principle of fairness within a dynamic intertemporal allocation of humanity's remaining carbon budget (based in turn on politically constructed temperature targets rather than harm avoidance) within an account of intergenerational justice. Such an account of intertemporal mitigation burden not only better captures the agency and responsibility issues of can-kicking, as the next section shall further explore, but it is also more normatively generative in its prescriptions.

Such a formulation theorizes mitigation obligations and therefore also can-kicking in terms of just shares of a just carbon budget, where individual or collective mitigation duties can be conceived in terms of shares in a carbon budget, which allocates the remaining emittable carbon among groups and their members as well as over time. A just allocation along either dimension would be one in which each agent did their part in a collective effort that satisfied imperatives of global as well as intergenerational justice. While such a duty of intergenerational distributive justice may prescribe a similar global emissions profile over time, compared to what would be required to avoid harmful climate disruption, the focus on mitigation burden-sharing avoids the thicket of person-affecting objections associated with the Non-Identity Problem that would plague any intergenerational climate ethics that relies upon imperatives to avoid causing either direct or indirect (via disjunctive wrongs that are defined in terms of ongoing capacities to avoid causing direct harm) imposition of harm. Under such a burden-sharing approach, the intergenerational carbon budget would first determine the total allowable emissions within a given time period (like a year), a subsequent allocation would allot shares of that total, first to collectives like nation-states and then within such collectives at least hypothetically also to individual members.

Despite the numerous conceptual and practical difficulties in prescribing individual carbon allowances, including the ethical indefensibility of viewing individual mitigation efforts as alternatives to collective and political actions, one can at least posit that in principle an individual person's share of this carbon budget would reflect both dimensions of climate justice. As shall be considered further below, individual members of collectives like nation-states may not be morally required to adhere to just personal shares of their national carbon budget if this is conceived as a mimicking duty to be followed only under conditions of collective can-kicking, but states may assign their members just personal carbon allowances and these may in some cases be sensitive to whether and how many others adhere to their personal allowances.[8] At any rate, can-kicking may here be provisionally defined in terms of personal or collective shares of a just carbon budget, rather than by actions that do or do not contribute to harmful climate impacts.

Agency and Responsibility in Can-Kicking

Whether in diagnosing fault or prescribing remedial actions, an ethics of can-kicking needs a cogent account of agency and responsibility. One can understand just and effective climate mitigation—and therefore also failures to take morally adequate action—as involving both individual and collective forms of agency, including interactions between the two. Persons do make choices that result in wide variation among individual carbon footprints, with some such footprints adhering to carbon budget limits and others exceeding them. While many factors that feed into an individual person's carbon footprint are determined by the collective agency of their larger society over which individual members typically have little or no control—the zoning and urban design decisions that affect their daily commute, the composition of energy portfolios, the availability of transit options other than automobiles, applicable regulations on industry and agriculture, their shared social emissions like those from governments or militaries, social norms around pollution and waste—some contributing factors to personal footprints are subject to personal choices within those social constraints and thus depend in part on individual agency.

Other sources of greenhouse emissions cannot be directly attributed to individual choices and so are often characterized as following from the collective agency of states, whether the result of policies or regulations governing fossil fuel use or greenhouse gas pollution or from structural features of energy and transportation infrastructure, building design and urban planning, or the availability of renewable energy resources. States are also treated as collective agents under the international climate policy architecture of the UNFCCC, since only states or governments can exercise the sovereignty necessary for entry into such treaties. Acting as agents on behalf of nations or peoples, states are routinely assigned responsibilities under international law and are liable under such law for their violations. While substate actors also routinely pledge mitigation actions under the rubric of the Paris Agreement, only states are required to make commitments under that convention. Even though these "nationally determined contributions" to mitigation (or NDCs) are not legally binding, their formal recognition connotes a moral and legal standing for states that requires collective forms of agency and responsibility.

These two forms of agency and responsibility can and do interact, and an ethical analysis of can-kicking should be prepared to articulate how they do. One can conceive of and should be able to articulate how and when they might diverge, resulting in collective can-kicking where some nation-state fails to meet its national obligations but where individual members of that state could not be held vicariously liable for their role in their polity's failure. To the extent that states or governments are responsive to their constituent members, one should be able to specify under what circumstances and to what extent individual members can be held responsible for the acts or omissions of states and governments and when they cannot, as well as what related duties they may have with respect to collectives to which they belong (termed *promotional duties* because related to individual roles in collective actions). These individual and collective duties should also be specified as to whether they apply under conditions of full or partial compliance, as in debates in climate ethics about slack-taking under partial compliance as well as being updatable over time and as conditions change or based on the actions or inactions of others.

Much of the dynamics noted above are more readily captured by burden-sharing accounts of mitigation duties in the climate justice literature rather than harm-based accounts from climate ethics approaches. Where collectives like nation-states fully comply with the fair mitigation burden-sharing targets that have been assigned to them, neither the collective as a whole nor any of its members would be committing morally egregious can-kicking. Burdens may have been inequitably allocated within the group, or some members may shirk their individual burdens without bringing the collective out of compliance, but the present cohort as a whole would be meeting its obligation to future cohorts rather than imposing additional burdens on them by deferring adequate actions, so no cans would be kicked. But dynamics between individual and collective responsibility do arise when one or the other engages in mitigation can-kicking.

Where their collectives fail to undertake adequate mitigation actions (i.e., under non-ideal conditions of partial compliance), climate ethicists disagree about what this entails for individual duties as well as about whether and the extent to which members share responsibility for such failures with their collectives. While noting the primacy of promotional duties whereby persons are obligated to support collective and political mitigation actions, some do urge or allow for individual mitigation actions beyond what is socially required for members of nation-states that fail to fully comply with their collective mitigation burdens. Elizabeth Cripps, for example, notes that there could be a "clean-hands" case for "mimicking duties" (defined as duties "to do what would be one's duty in some fairly organized collective response")[9] where "it is genuinely impossible to fulfill promotional or direct duties" as well as "a fairness-based case for mimicking when compatible with them," but she argues that such duties can never be exclusive or primary.[10] Similarly, John Broome argues that the most effective forms of individual action are "through political action to induce your government to do what it should" but that "reducing your carbon footprint to zero may contribute indirectly to that effort."[11] For both Cripps and Broome, our primary duty would be to establish just institutions and policies related to climate mitigation, which in this context would require our compliance with their terms once established, not to take mitigation actions outside of the context of their potential instrumental role in this process. With individual members tasked primarily with promotional duties to urge collective and political mitigation actions rather than undertaking their share of such actions individually, can-kicking would appear to be largely if not exclusively a collective phenomenon, with individuals tasked only with urging their polity's compliance with just collective obligations rather than duties to take those on individually. Harm-based accounts equivocate on whether or what mitigation actions individual persons should undertake now, while social institutions of climate justice remain elusive, given difficulties in linking those individual actions to either any discernable climate impact or the establishment of just institutions. Justice-based accounts offer a clearer and more compelling account of why individual persons ought to undertake mitigation actions now *while also* working to establish political institutions requiring others to also do so (and even when other societies have not yet done so). Rather than only being normatively limited to specifying mitigation duties once a fully global institution of climate justice has been established, the justice-based account can furnish normativity to mitigation actions during the (necessary and perhaps permanent) transition.

A related question about the dynamic between the duties of different agents asks whether those complying with their justly assigned mitigation duties can become responsible for "picking up the slack" when others fail to comply, kicking their cans down the road. David Miller, for example, argues for a slack-taking duty in cases where victims can make demands of justice for rescue, which would seem to apply to climate mitigation duties. "However much we might regret the non-compliance and condemn the non-compliers," he writes, "this is simply a fact of life, so what we must do now is to recalculate contributions, distributing responsibilities fairly among the coalition of the willing, so to speak."[12] While individual agents need not take up the slack left by the non-compliance of collectives (as this would be excessively burdensome and probably impossible in most cases, nor must they engage in mimicking duties on such occasions), and some object to slack-taking duties altogether, slack-taking may be required in cases of can-kicking among either collective or individual agents. Insofar as taking up slack with additional mitigation efforts beyond what was originally assigned may be required in such cases, processes for making adjustments to initially assigned mitigations become necessary. This dynamic thus necessitates some means of periodic review and reallocation, characterized here as *taking stock*.

Taking Stock

Like the Global Stocktake,[13] which under the Paris Agreement requires that parties to the convention meet every five years to review progress toward international climate goals and revise their contributions to this collective effort, taking stock in the sense relevant to mitigation duties requires periodic revision of carbon budgets and perhaps even of the objectives on which such budgets are based in light of mitigation actions and failures since the last budget allocation. If, for example, the carbon budget allocation for the United States of humanity's final half trillion tons of carbon was 20 billion tons by 2050 (the international allocation),[14] and the share of that for the period 2021–25 was four billion (the intertemporal allocation) as part of a drawdown that would be followed by three billion tons in 2026–30, then stock would need to be taken toward the end of that first budgeting compliance period and in advance of the second. If only revisiting progress toward the intertemporal allocation, expected failure to limit carbon emissions during that first period might require a steeper drawdown in the second—say, 5.5 rather than 4 billion tons in the first might require 2.5 rather than 3 billion tons in the second in order to correct course to remain within the budget of 20 billion tons by 2050. Of course, stock-taking of this sort would require revision not only of the carbon budget targets themselves but also of mitigation actions undertaken in pursuit of those targets (and likely also the magnitude of social investment in mitigation), as the targets are designed to inform the actions and failure to meet a target entails that ambition in mitigation actions accordingly be increased.

The international allocation might also be revisited, with the US share (for example) revised upward or downward from that 20-billion-ton budget, whether due to new information about links between carbon emissions and temperature targets coming to light or from revision of temperature targets themselves, from changes in international balances of power or populations or other bases of national carbon allowances, or because of changes in available technologies or other conditions that affect carbon budgeting assumptions under the initial agreement. Perhaps the US share of the total 2050 carbon budget might increase to 24 billion or decrease to 16 billion tons from this stocktake, requiring revision to the slope of that drawdown curve during the next compliance period. Or perhaps the US budget could be increased in order to take up the slack from other states failing to meet their carbon budget targets, as considered above.

Stock-taking at the collective level can also reflect several uncertainties that elsewhere serve as objections to strong intergenerational duties and incorporate these into revised budgets. If some new zero-carbon energy technology were created that is capable of being deployed at scale and for relatively low costs, such that per-unit mitigation costs could rapidly be brought down without proportionally rapid abatement cost increases, then the potential gains from efficient technology could inform future carbon budgets. Likewise, if one or more of the state parties that are under carbon budget mitigation obligations suddenly becomes very rich or very poor, affecting the formula for determining their equitable carbon shares, this could also be incorporated into future compliance rounds. Likewise, changes to climate impact forecasts, adaptive capacity, climate engineering technologies that alter either overall carbon budgets or their allocation among various parties, or other similar developments could also feed into ongoing mitigation efforts through revision of either the budgets themselves or the means of realizing their targets. Rather than allowing such considerations to undermine the cogency of intergenerational obligations, the process of stock-taking could accommodate their critical force.

Notionally, at least, this budget and drawdown curve could be disaggregated to the level of individual carbon allowances for all residents of the US and for each month of every year within a compliance period. While such allowances would of course need to account for social emissions that don't accrue to specific persons, accommodate new members of the polity and reflect relevant differences among the members that might affect their just personal allowances, and address several other contentious issues in carbon accounting (e.g., between production-based and consumption-based accounting), one could at least in principle posit that there could be a just monthly carbon budget for every US resident that informs their individual mitigation burdens and perhaps guides their mimicking actions. Stock would then need to be periodically taken of these, too, and beyond the kinds of considerations that inform the international and intertemporal carbon budget allocations described above. This stock-taking exercise could potentially illuminate the connections between hyperlocal and global, between contemporaries and future persons, and between private and public political actions that theorizing climate responsibility in its individual and collective dimensions is intended to capture.

At the individual level, this stock-taking could for example include consideration of the value of public investment in energy or transportation infrastructure that would allow persons to more efficiently meet their emissions budget targets. For example, those residing in areas with poor transit options other than private automobiles might take stock of the concentrated carbon budget impact of private automobile commuting and discover that by carpooling they could more easily meet their individual mitigation obligations or could even compare the costs in increased taxes for their urban area against the reduced transport emissions resulting from supporting such social investment. Those failing to meet their carbon budget targets and needing to more rapidly decarbonize could increase the slope of their drawdown curves and perhaps more readily meet the targets of their next compliance period through other cooperative or private but socially beneficial actions designed to reduce their personal carbon footprints, from eating lower on the food chain to installing rooftop solar panels or replacing their inefficient kitchen or HVAC appliances with more efficient alternatives. In taking stock of their personal progress toward meeting carbon budget targets, they could not only refine their targets but also make informed choices among actions designed to help them meet it. To the extent that they can consider social investment in public energy or transportation systems as well as private investment or behavioral changes, this stock-taking exercise can highlight a common purpose behind the low-carbon transitions undertaken at a community level,[15] many of which are likely to provide more efficient returns than are available through private mitigation actions while also revealing the low-hanging fruit of private mitigation options.

Varieties of Can-Kicking

On the basis of this formulation of can-kicking in terms of the allocation of mitigation burdens through assigned shares of a carbon budget, one can differentiate between several distinct varieties of can-kicking. First is the most egregious: rejecting all carbon budgeting constraints altogether. Whether from denial of causal links between greenhouse emissions and climate impacts, indifference to the suffering of others, or unfounded optimism about geoengineering solutions or adaptive capacities, refusing constraints in this way willfully exposes others to harm and leaves future persons with correspondingly greater mitigation and adaptation burdens, taking no responsibility for the harmful impacts of current actions on contemporaries as well as future generations. Such outright refusals to acknowledge the causal linkages between current actions and future harm strain the metaphor of can-kicking, since they would seem to deny the existence of any can to kick but are often publicly justified in terms of delay and deferred actions (often with the call for further study before undertaking any action but with the intention of merely postponing imperatives to act rather than improving knowledge).

Less egregious but still morally problematic is can-kicking that recognizes and accepts global mitigation imperatives but refuses to adhere to equitable carbon budget limits. As part of its "nationally determined contributions" to global mitigation under the Paris Agreement, for example, the US might claim to be entitled to 30 rather than 20 billion tons, say by pledging 30 billion in its NDC, adhering to such limits in its mitigation plans but in doing so causing the overall carbon budget to be exceeded and thus also the associated temperature target to be missed. Alternatively, it might accept the climate budget target but then pledge reductions that fail to meet it (say, by publicly acknowledging a budget of 20 billion tons but then pledge mitigation efforts that result in 30 billion, kicking cans in its linking pledges to budgets) or accept and pledge 20 billion but then implement mitigation programs that fail to meet pledge targets (kicking cans in implementation of pledge commitments). Under the Trump administration, the US engaged in the first kind of can-kicking, openly denying climate science and rejecting all limits on greenhouse emissions as well as the authority over domestic policy of international institutions like the UNFCCC. Under the Obama and Biden administrations its can-kicking was of the second and third kinds, endorsing the authority of the UNFCCC as well as the need for domestic mitigation actions but declining to accept equitable carbon budget targets through its NDCs (which collectively fell far short of what would be needed to adhere to the higher 2 degree temperature target endorsed through the Paris Agreement) and also failing to put into place domestic mitigation programs sufficient to meet those pledges (with the Clean Power Plan, which served as the Obama administration's key policy mechanism for meeting its NDC pledges in Paris, never taking effect). In doing so, those in command of US policy have shifted mitigation burdens and/or adaptation costs onto future persons, including younger contemporaries as well as those not yet born.

Can-kicking may by contrast merely reflect mistaken impact estimate or other forecasts, where planned mitigation actions either had less decarbonization impact than expected or were undermined by emissions increases elsewhere. Honest mistakes in such estimates do not really constitute wrongful imposition of burdens onto future others so much as being disappointing investment yields, and need not be regarded as unjust so long as they get corrected in future stock-taking assessments. Where can-kicking follows a drawdown curve that begins modestly but includes realistic plans to meet carbon budget constraints in the intermediate future—as with pledges for carbon neutrality by 2050 made in 2024 that defer most actions to the 2030s and 2040s (when current executives or politicians are no longer in office)—the temporal burden-sharing allocation appears to be morally dubious but could at least in principle meet the demands of climate justice over time. If, for example, deferring steep cuts in emissions was for the purpose of developing and deploying new energy or transportation infrastructure that takes years to be brought online, the result of transitions in technology or other decarbonization efforts that entailed a significant lead time before they began to take effect but reflected realistic mitigation potentials, it may also not be impugned as ethically negligent can-kicking.

Discerning the difference between sincere mitigation efforts that come with deferred impacts and mere can-kicking can be difficult, as disingenuous plans to merely deflect and delay must appear to be sincere in order to effectively deceive. Often key to determining whether some set of announced mitigation plans are of the former kind are near-term authorizations and investments. Significant investments in research and development of new technologies or systems to shorten the curve for making them scalable or deployable may entail deferred benefits but do not defer actions or costs.

On the other hand, mere pledges of actions or results beyond election cycles or quarterly reports in the absence of palpable near-term commitments may be designed to secure reputational gains in the present while deferring the costs needed to support them onto others. In declining to shoulder any near-term burden—whether in the form of financial investment, political capital, social or intellectual resource deployment, or mobilizing any other change-inducing resource—pledges of intermediate-term future action or promises of deferred results are likely to amount to morally dubious can-kicking (a finding that may only be able to be confirmed when the pledged actions or promised results fail to obtain).

Likewise with individuals, our assessment of can-kicking depends on whether personal mitigation actions are merely being deferred to a later time or are being shirked altogether, and thus transferred and imposed onto others. If, for example, I have a personal carbon allowance for a calendar year that represents my just share of a carbon budget and I emit 90% of that allowance over the first two months, my options are limited from March through December. Either I could dramatically reduce my emissions for those remaining ten months in order to remain within my budget, in which case my can-kicking may have been imprudent but was not wrong or unjust, or I could simply allow myself to go over budget, claiming more than my just share of the budget and burdening others with my excess. In either case I would have kicked the can along during January and February, with the difference being between having to bear the costs of my deferred action myself and imposing that cost onto others. To whom we kick the can matters. Conceptualizing our individual roles in sustainable transitions (along with our civic obligations to engage in promotional actions) in terms of budgets, drawdown curves, and an ethics of can-kicking may introduce constructive forms of reflexivity into our consumer choices and strengthen our perceived connections between self- and public interests.

Conclusion

I have argued for a conceptualization of "kicking the can down the road" in shirking or deferring of individual and collective mitigation duties in terms of just shares of a carbon budget that is justly allocated over time and then among contemporary collectives, then finally (and notionally) also to individual persons. Without being able to specify what we *ought* to do in this domain of climate ethics, we cannot say whether or not individual or collective agents have satisfied their mitigation duties or whether they have kicked that can down the road. When cast as a duty of avoiding harm, in a context where such avoidance is no longer feasible, can-kicking appears as inevitable and its avoidance impossible. However, refocusing imperatives to mitigate ongoing contributions to climate change as involving participation in a cooperative effort guided by a principle of fairness in burden-sharing can accommodate politically constructed temperature targets as feasible and legitimate objectives while allowing carbon budget allocations to specify mitigation duties while recognizing the roles of both individual and collective agents and linking their responsibilities to cooperatively act. Grounding this account of can-kicking ethics in fair mitigation burden-sharing rather than person-affecting harm also avoids the trapping of common objections to intergenerational obligations like the Non-Identity Problem, while including a process of periodic stock-taking adds an element of reflexivity to ethical deliberation and its conversion into imperatives of action while accommodating other common objections to intergenerational accounts of obligation. So long as humans remain inclined to kick this can down the road, as we seem inclined to do, we should at least acknowledge what we're doing.

Notes

1 Article 3, Principle 1, United Nations Framework Convention on Climate Change.

2 For example, see James Hanson, *Storms of My Grandchildren: The Truth About the Coming Climate Catastrophe and Our Last Chance to Save Humanity* (New York: Bloomsbury Press, 2009), and Mary Robinson, *Climate Justice: Hope, Resilience, and the Fight for a Sustainable Future* (New York: Bloomsbury Press, 2018).

3 First developed by Derek Parfit in *Reasons and Persons* (Oxford: Oxford University Press, 1984), the Non-Identity Problem challenges the premise that actions undertaken in the present can harm specific persons in the future insofar as those actions can also affect the identities of persons who will come to be born in the future. For any possible set of future

persons, then, our present actions cannot be said to make them worse off, since different actions (harmful to the environment though they may be) would result in different persons being born in the future. The problem has become a key challenge for intergenerational ethics.

4 "Causes and Effects of Climate Change," United Nations, www.un.org.

5 Lucas Stanczyk, "On the Moral Challenge of the Climate Crisis," in this volume.

6 Stanczyk, "On the Moral Challenge of the Climate Crisis."

7 Aaron Finley, "Slack Taking and Burden Dumping: Fair Cost Sharing in Duties to Rescue," *Journal of Ethics and Social Philosophy* 23, no. 3 (2023): 343–364, at 343.

8 See Keith Hyams, "A Just Response to Climate Change: Personal Carbon Allowances and the Normal-Functioning Approach," *Journal of Social Philosophy* 40, no. 2 (2009): 237–256.

9 Elizabeth Cripps, *Climate Change and the Moral Agent* (New York: Oxford University Press, 2013), 22.

10 Cripps, *Climate Change and the Moral Agent*, 138.

11 John Broome, *Climate Matters: Ethics in a Warming World* (New York: W.W. Norton, 2012), 81.

12 David Miller, "Taking Up the Slack? Responsibility and Justice in Situations of Partial Compliance," in *Responsibility and Distributive Justice*, ed. Carl Knight and Zofia Stemplowska (New York: Oxford University Press), 234.

13 "Global Stocktake," United Nations Climate Change, www.unfccc.int.

14 Henry Shue, "Human Rights, Climate Change, and the Trillionth Ton," in *The Ethics of Global Climate Change*, ed. Doug Arnold (New York: Cambridge University Press, 2012), 292–314.

15 David Fleming, *Energy and the Common Purpose: Descending the Energy Staircase with Tradable Energy Quotas* (London: The Lean Economy Connection, 2007).

9

HARD TRUTHS IN CLIMATE POLICY AND POLITICS

SHELLEY WELTON

In his chapter in this volume, "On the Moral Challenge of the Climate Crisis," Lucas Stanczyk offers up some hard truths about the necessary response to the planetary emergency. He focuses on two challenges underappreciated by policymakers busy architecting industrial policy to "solve" the crisis: (1) there is no way to obtain the necessary scale and pace of emissions reductions without focusing explicitly on contemporary patterns of consumption; and (2) if left unchecked, global population growth threatens to overwhelm any gains otherwise made.[1] Stanczyk proceeds to develop a novel theory of intergenerational justice to guide us in the question of *how* we might tackle these thorny challenges, even as he asserts admirable humility about the precise answers.[2]

These are indeed *hard* issues. I have elsewhere described consumption as the "third rail" of climate policy,[3] and those who bring up population in the climate conversation are frequently branded as neo-Malthusians or worse.[4] Because these issues are such lightning rods, I begin this chapter by exploring the question: Are these really *truths*? Ultimately, I agree with Stanczyk that the climate crisis is unavertable without more attention to consumption and, in a way, population—but I think a more nuanced diagnosis of why and how each presents a critical "wedge" in the climate change solution pie chart clarifies his argument.[5]

After making these additions, I offer a policy scholar's reaction to Stanczyk's creative theory of intergenerational justice, endorsing it on slightly different grounds than he offers. Then, accepting the framework he lays down, I focus on some of the practical challenges that Stanczyk's thought-provoking theory raises. In doing so, I grapple with a hard question of my own: What role should climate ethics play in on-the-ground climate policy and politics? Stanczyk peppers muted but potent criticism throughout his piece of initiatives like the Green New Deal and of climate policymakers' failures to center hard truths about necessary lifestyle changes.[6] I have been struggling myself with the absence of what I consider many critical components of "good" climate policy from the much-lauded Inflation Reduction Act of 2022—the United States' most ambitious climate legislation to date.[7] Yet "politics is the art of the possible,"[8] and I am far from convinced that broadly invoking consumption- and population-related concerns in climate discourse and climate policy would move us forward, not backward, in our response to the problem.

Toward the end of his chapter, Stanczyk offers a brilliant cautionary intervention to champions of industrial policy focused on growing carbon-stanching industries such as carbon capture and storage and solar geoengineering. He warns that all of these technological gains might be usurped by those in power to prolong the use of fossil fuels, rather than transform the system.[9] I contend that this warning contains the seeds of wise guidance toward morally and politically sensitive climate policymaking: Policymakers must strive to enact policies that improve not just the technological terrain but the political conditions for more transformative system change. Our individual ethical duty, in turn, is not merely to tend our own climate garden but to organize toward such change.

Are These Truths?

Stanczyk's first hard truth is that we must face the need for real limits on consumption to respond with necessary speed to the climate crisis.[10] He is almost certainly right—indeed, even the Intergovernmental Panel on Climate Change (IPCC) finally included a full chapter on "Demand, Services, and Social Aspects of Mitigation" in its last assessment report.[11] Because we are in strong agreement, I wish only to refine and expand his claims on this point.

First, a refinement: Both Stanczyk's core point and the difficulty of the task he raises are punctuated by more granularity about *whose* consumption must change. After all, average per capita emissions in sub-Saharan Africa are already lower than what scientists calculate as necessary to reach net zero.[12] Even in the United States, per capita emissions among the poorest half of the population have declined since 1990.[13] As economist Lucas Chancel emphasizes, patterns of consumption-related carbon emissions are now largely explained by within-country inequality—a fact that both offers up a new set of solutions for consumption-related emissions and creates its own political challenges.[14]

Stanczyk appears to favor an approach that differentiates luxury emissions from others, astutely noting that "preferences" are not "all on a moral par."[15] I want to punctuate this distinction in types of emissions as a promising way to make progress in the difficult moral and political terrain of regulating consumption. Luxury emissions are often cast in opposition to "decent-living emissions," which Lukas Tank describes as those "that allow the kind of access to 'food, shelter, safe water and sanitation, health care, education, transportation, clothing, refrigeration, television and mobile phones' that is necessary for a minimally decent life."[16]

This distinction is critical in looking beyond raw emissions numbers to figure out *what* consumption to target and why. An illustration helps drive this point home. I'm currently part of a research project examining energy poverty in the southeastern United States. Part of this effort involves interviewing households that self-identify as energy vulnerable, and one such interviewee is a rural eastern Tennessean whom I will call "Jerry." Jerry is relatively sophisticated in his understanding of the energy system, with an excellent grasp of energy sources, major energy bill drivers, and energy management techniques. He explained to us that to keep his energy bill affordable, he powers only two things with electricity in his home: his refrigerator and a single overhead bulb. He goes without air conditioning. For heating, he scavenges abandoned coal mine sites for leftover chunks of coal to burn in a coal stove.

This profile suggests that Jerry is both struggling to get by and a fairly heavy consumer of carbon.[17] Any policy targeting absolute amounts of consumption-related carbon emissions might penalize him. Yet Jerry's carbon emissions are necessary for a minimally decent life under the structural constraints that he faces. He would not burn coal if he could afford electric heating. The electricity he does consume comes from a power grid in the southeastern United States, making it carbon-intensive by nature. Jerry has no realistic, affordable options for managing these emissions.

In the long run, the hope is that the entire global population will be able to construct decent lives without burning carbon through a widespread transition to clean energy sources.[18] But in the meantime, Jerry is not culpable for scraping by until larger, structural changes make this shift possible.[19] Put differently, the carbon emissions associated with achieving decent living standards should not be the target of consumption-specific policies. Instead, they should be the grist for production-side reforms. In contrast, luxury emitters have no moral claim to the remaining carbon budget, even as they have ample ability to make consumptive choices that limit their emissions.[20] Defining what counts as "luxury" emissions and focusing consumption-related climate policy on these goods and services is thus a moral and fruitful path for advancing Stanczyk's appropriate concerns about carbon emissions from consumption.

But of course, on a more practical level and as Stanczyk well realizes, luxury emissions are correlated with the very wealthy and powerful who currently dominate politics and naturally resist such controls. Instead, they offer techno-optimistic solutions to the climate crisis that will require no sacrifices on their part.[21] Stanczyk could take on this group and their suppositions more directly in making his case for limiting consumption. Techno-optimists argue that present cuts to fossil fuel–based consumption are ill-advised because they risk slowing technological progress toward solutions that can wholesale replace fossil fuels without significant lifestyle disruptions. For the techno-optimists, a panoply of new technologies is the magical hangover elixir that will cure our carbon binge. Indeed, such technologies may be critical to a 1.5- or 2-degree pathway: IPCC models project carbon removal and carbon capture and storage playing *enormous* roles in counterbalancing continuing emissions in coming decades—despite their nascent status.[22] Similarly, there are news reports every few years of a plane flown on solar power or kitchen scraps, and episodic fascination with meat grown in labs.[23]

These optimists have a historical leg to stand on: Doomsayers that have long predicted that the Earth cannot possibly support our expanding lifestyles have been proven wrong many a time, at least on the timescales they projected.[24] I think there is a convincing argument that climate change is different, but there is a strong human—or perhaps just capitalist—impulse to insist it is not. Indeed, entire countries' and corporations' plans for decarbonization are built on this impulse: Southern Company, one of the largest US utilities, is transitioning all of its coal generators to natural gas this decade, banking (or assertedly banking)[25] on forthcoming carbon capture and storage to allow it to reach net zero by 2050.[26] Switzerland has been explicit about its need to rely on both technological carbon dioxide removal and investments in actions in other countries to reach its targets.[27]

Relying on this type of magical hangover elixir in mitigation planning on the timescale of generations is not just ill-advised but unethical, under the framework Stanczyk builds. If he were to tackle head-on the relationship between present-day luxury consumption and techno-optimism in intergenerational climate policy, he might punch his point home further. In the face of not knowing whether these tools will scale, cuts in certain types of consumption are morally demanded. The long line of work on the precautionary principle could prove useful here.[28]

Stanczyk's second hard truth is that "we cannot think responsibly about the climate crisis without evaluating . . . global population size."[29] I have more hesitations about this part of his analysis. There is much truth to Stanczyk's observation that policymakers have shied away from this topic—but much dissension as to why. Several scholars continue to emphasize population growth as a core contributor to climate change.[30] Others—notably many feminist scholars and scholars of global environmental justice—have mounted sustained attacks on the centering of population in climate conversations.[31]

Stanczyk doesn't give us much in terms of his thinking regarding how best to intervene in this fraught debate. His justification for enhancing the focus on population is rooted in aggregate numbers: "With every additional 500 million people on the planet, global greenhouse gas emissions are expected to grow by at least one billion tons of carbon per annum." Ergo, reduce population size, reduce emissions.[32] To concretize his reasoning, Stanczyk provides us the example of a couple with two children, deciding whether to have a third child while living on a plot of land that can only feed four. He suggests that it would be unethical (under the intergenerational framework of justice that he develops—more on this later) for the family to have a third child, because it would mean inadequate nutrition for the existing children.

This is obviously and intentionally a stylized hypothetical. But place this family down on Earth and the picture gets far more complex. It turns out, there is plenty of food to feed this hypothetical third child, but it's rotting in US (and other) food distribution channels rather than getting sent where needed.[33] Thus zoomed out, the family no longer looks like the unethical actor.

Now, unlike in the food context, and as aptly discussed by Stanczyk, we *do* face a real shortage of remaining tons of carbon in our budget.[34] But from a carbon perspective, again, it is choices about the size of high-wealth populations that matter most, because population growth's impact on climate is quite clearly intermediated by affluence.[35] Adding one person in the United States (per capita CO_2 emissions of 14.86 tons) is, on average, about *five hundred times worse* than adding one person in the Democratic Republic of Congo (per capita CO_2 emissions of 0.03 tons).[36] And adding a billionaire baby is really the worst of the worst—except, of course, that it's counter to many of the core tenets of the reproductive justice movement to think in these crude demographic terms at all.[37]

If this diagnosis is correct, then there is still an important, complex conversation to have about the convergence of population and the climate crisis—it's just not the one that is often front and center regarding "high-fertility" but low-emissions countries.[38] In addition to its colonial overtones, there is a practical reason that a focus on these countries is misplaced: Population and consumption are not independent variables. Demographers have long recognized that economic development leads to a significant drop in fertility rates.[39] This pattern has persisted across continents and cultures that have experienced rising affluence, to the extent that many developed countries now have birth rates *below* replacement rates—causing their leaders to identify precisely the opposite demographic crisis from Stanczyk.[40] It is thus likely that increases in per capita carbon emissions that accompany the development of less developed countries would *naturally* lower birth rates. If that is so, then either (a) high-fertility, low-emissions countries do not develop and their population growth remains peripheral to addressing the climate crisis; or (b) high-fertility, low-emissions countries do develop and their fertility rates naturally fall, in line with long- and widely-observed trends.[41] Obviously, the second case is far preferable to the first from a moral perspective, assuming that development comes with increased human flourishing. But critically for present purposes, in either case, a focus on controlling population *as such* in these countries may not dramatically affect emissions. Indeed, some research suggests that lowering fertility in underdeveloped countries might itself *induce* development, such that more focus on population might occasion *more* climate emissions, not fewer.[42] All to say, the relationship is complex.[43]

What is less complex for purposes of climate change is the pursuit of appropriate population-adjacent policies in countries that are far above average global per capita emissions levels. The most impactful, short-term way to influence carbon emissions via population might be to focus on US (and other high-consumption countries') reproductive justice policies—including, quite glaringly, recent contractions in the availability of reproductive autonomy for childbearing people in this country.[44] A second important area of focus is rethinking burgeoning nativist policies aimed at increasing domestic fertility rates in developed countries.[45] Countries panicking over below-replacement fertility levels have numerous options other than increasing birth rates. These might include reconsidering core economic policies that require a ballooning workforce and reimagining the economics and politics of care work. Moreover, expanded immigration quotas in high-emissions countries—particularly if targeted toward accepting emigrants from climate-ravaged areas—might prove an ethical and climate-responsive alternative to fertility-enhancing policies.[46] In his evocative discussion of the social consequences of climate disasters, Stanczyk implicitly recognizes that the morality of our immigration policy and our climate policy are deeply linked.[47] This linkage, to me, should form a core of any discussion of climate and "population" policy.

To be clear, Stanczyk may well agree with all I've said here, but he does not wade into particularities in his piece. Instead, in a footnote, Stanczyk nods to the horrors that have attended policymakers' attention to overpopulation in a world rife with racism, xenophobia, and fascism. But, he asserts, that is no reason for not at least identifying the moral imperative to *do something* about it—just not what has been done in the past.[48] This offhand acknowledgment is inadequate, by my lights. Sounding generalist alarms about the need to confront "global population" as an ethical imperative—an imperative that is more contingent than he states—may do more harm than good, even as a jumping off point.

Intergenerational Justice, One Generation at a Time

One of Stanczyk's core contributions is a novel version of intergenerational justice that he derives by rejecting utilitarianism and focusing instead on satisfying the Non-Identity Problem in the realm of climate change. Stanczyk convincingly argues that utilitarianism provides an illogical basis for reasoning about duties owed to future generations, given its focus on summed utility.[49] He then suggests that a non-utilitarian theory of intergenerational climate justice requires overcoming the challenge that any interventions made to improve the climate will produce a different set of human beings than would have existed under climate-destructive policies. Consequently, we cannot identify with precision any person yet to be born who will be specifically harmed by failing to act on climate change.[50] Stanczyk's solution to this dilemma is to propose a human-lifespan theory of climate justice that demands that "institutions must be erected that continuously give every living person's diverse claims the appropriate weight as soon as and for as long as they live."[51] Over time, he asserts, this theory will generate adequate moral guidance so long as every generation makes it possible for all people living to avoid both present and future harm.[52]

As a disciplinary outsider, the non-identity line of reasoning always takes me by surprise. With a problem like climate change, which is likely to make *every future person's* life worse, I am not sure it takes being a committed utilitarian to believe that intervention is morally required despite its inevitable alteration of precisely who is born. It seems enough to me to reason that climate intervention is (at least in part) about handing each non-identifiable but actual future person a life as dignified, meaningful, and free from suffering as we can manage.[53]

I also have some more applied inquiries about the theory: What if the world finds out that the main climate tipping point will suddenly occur in two hundred years on current emissions trajectories? Does Stanczyk's theory leave the present generation without a duty to forestall this, because no contemporaries will themselves experience the tipping point or likely be alive with those who will? Are we okay with that? In contemplating this (also stylized) example, it strikes me there may be reasons for holding onto a longer time horizon, such as the commonly cited "seven generations" paradigm of Iroquois natural resource management.[54]

That said, I find Stanczyk's theory quite appealing in the climate context for a different reason—one that he emphasizes later in the chapter. That reason is the immense uncertainty surrounding both climate consequences and climate interventions. In fact, there is much we don't know about the magnitude or timing of climate tipping points, what social and political reactions to them might be, or how and when technological and social landscapes might shift.[55] Stanczyk's contemporary-regarding theory works well in these conditions, demanding precaution but not prescience.

POLICYMAKERS' DUTIES—AND OUR OWN

As a philosopher, Stanczyk is right to center hard issues and grapple with how theories of justice might help us approach them—and to call for more transdisciplinary research into the same. But part of his critique is that climate *policymakers* are "systematically avoid[ing]" these issues in "mainstream policy discussions."[56] In this final part, I work through this claim to consider what exactly the duty of those attempting to craft or influence climate policy might be, under his intergenerational theory. As he suggests, I think the answer is quite difficult—and might well point in a different direction than more overt discussion of consumption and population.

Stanczyk argues that policymakers have a duty to tackle both these topics in climate policy.[57] But to what end? I suspect that any policy that restricted Jerry's ability to glean for coal and forced him to buy unaffordable electric heat, or increased the price of staple but high-carbon products such as meat, would prove wildly unpopular with him and others in similar situations. This intuition forms the basis of geographer Matt Huber's argument in *Climate Change as Class War*. Huber contends that the obsession with outsized individual consumption is really a professional class distraction that resonates not at all with lower- and middle-class Americans, who have experienced shrinking spending power over the past several decades (along with falling emissions).[58] Consequently, he suggests, focusing on changing the consumption habits of the masses is simply bad politics, unlikely to garner results.

If everyday people seem to want to ignore hard truths, US congresspeople appear far worse—influenced not only by their constituents' desires but also the financial imperatives of running for office and their own elite consumption preferences. And under hyperpartisan conditions, it is the raw numbers of the US Congress, along with the convoluted rules of reconciliation, that ultimately dictate the shape of US climate legislation. These dynamics caused the 2022 Inflation Reduction Act (IRA) to focus predominantly on industrial policy to ramp up domestic production of core climate technologies, from renewable energy to hydrogen, nuclear, carbon capture and storage, and carbon dioxide removal.[59] These investments have since been partially repealed at the urging of the second Trump administration, and even in their most robust form were far from enough to put the US on the necessary path to net-zero emissions.[60] But the IRA would not have been more robust had policymakers pushed to center the themes of consumption and population—if anything, it might have never gotten off the ground. So, were policymakers unethical to background these issues? Or were they morally justified in doing so, to gain the climate progress that they did?

Stanczyk offers an essential warning to those who would reflexively justify these political machinations. As he explains, climate will create "auto-catalytic social effects," as disasters disrupt communities and "provide the occasion for yet more wrongdoing by powerful people bent on exploiting all available fossil fuels."[61] If industrial policy interventions do nothing to change political and social conditions, then they might be seized upon to *expand* the production of fossil fuels, not wind them down.

This point is crucial and disheartening, especially if one thinks there is nothing to do about it. Stanczyk appears resigned about the immutability of elite domination, wryly observing that "the powerful will predictably not obey any of the most important imperatives of climate ethics."[62] Left unexplored are the linkages that undergird this assumption—linkages among ethics, policy, law, power, politics, domination, and resistance.

Those of us who spend our time in these domains do so because, in spite of the odds, we believe there is space here for transformation. Fortunately, exploring these connections forms the core of a (re)emerging movement in legal scholarship focused on the interrelationships of law, social order, and power. The Law and Political Economy movement, drawing from several predecessor movements, has resurrected questions of how and why the law serves as a tool to maintain and reinforce elite economic domination.[63] While too immense a movement to do justice to here, the point to emphasize for present purposes is this: Laws intermediate social and economic relations and have the power to shift them, even as these relations also help shape the content of law.[64] Any set of laws creates what Samuel Moyn describes as "situated freedom"—some room (sometimes less, sometimes more) within them in which "critique and transformation" can occur.[65]

I'm curious about how law's situated freedom might relate to Stanczyk's ethical imperative. To try to put some pieces together: Stanczyk's concern is that we are failing our intergenerational ethical duty by not addressing population, consumption, and other hard climate truths. My concern raised above is that centering these issues might actively set back climate policy in the current political order. But this political order is not immutable, and emerging insights from law and political economy offer us the possibility to interrogate law's potential role in the transformation of this elite, fossil fuel–dominated political order. Put these together and I think you have the beginnings of an ethical pathway forward for policymakers today.

If Stanczyk's diagnosis of the ignored hard truths of climate policy is correct, it is not enough for policymakers to simply "do industrial policy" and hope the system sorts out where to go next. They must actively theorize and embed within these laws possibilities for larger structural change. In other words, they must build in what situated freedom they can—and we must push them to do so.

Amna Akbar's powerful exploration of "non-reformist" reforms is a useful way to understand how law's situated freedom can manifest in legislative change. Akbar characterizes "non-reformist" reforms as those that "aim to undermine the prevailing political, economic, social order, construct an essentially different one, and build democratic power toward emancipatory horizons."[66] These types of reforms engage with systems where they are, but push them in directions that expand the possibilities for more radical dialogue and change in the future. In doing so, they "seek to redistribute power and reconstitute who governs and how."[67]

What might such non-reformist reforms look like, concretely, in climate change law? I want to draw in here a concept advanced by legal scholar Shalanda Baker, who excoriates the practice of "climate change fundamentalism." Baker defines "climate change fundamentalists" as those "activists who advocate for policies to mitigate and adapt to the impacts of climate change without concern for issues of equity."[68] Baker worries that a focus within climate policy on reducing tons of carbon alone, without attention to policies' distributive impacts, risks exacerbating structural inequalities and environmental injustice.[69] This worry is related to Stanczyk's warning that industrial policy alone might serve more as a cover than a cure for entrenched fossil fuel interests. Both evince an awareness that any climate change policy that is focused exclusively on technological decarbonization contains an explicit and risky choice *not* to recognize inextricable linkages among climate change, capitalist economic and social relations, inequality, colonialism, race, and broader ecological threats beyond mounting carbon.[70]

I want to posit that policymakers' duty in attempting to address climate change under real-world political conditions, in a way that respects intergenerational equity, is to resist climate change fundamentalism. Ethical climate policy must foreground—as much as politically possible—the linkages between climate change and the conditions that produce what often feels like intractable elite domination, which in turn produces the focus on material-consumption-as-wellbeing that Stanczyk insists must change.

I harbor more optimism about the broad program of the Green New Deal than Stanczyk does because I see it as an effort in this vein. In its fullest, "radical" form, the Green New Deal offers a vision for a new sort of abundance rooted in leisure time and enhanced social ties, rather than material consumption.[71] It gets there via significant shifts in labor power and job conditions, public investment and ownership, and government support for care work and workers, broadly defined.[72] Under this vision, a renewed public sector and a shift in social and economic power dynamics unlock the potential for more dramatic shifts in how we eat, play, work, and live together.

This program is, arguably, a more oblique but politically efficacious way of tackling Stanczyk's hard truths about the extent of social change necessary to adequately respond to climate change.

The United States obviously is not yet ready to accept this version of a Green New Deal.[73] But even the IRA—pared down as it was— contained seeds planted in this fertile ground. Although the Act doubled down on neoliberal private investments as its core climate strategy,[74] it had provisions to strengthen worker power through prevailing wage and apprenticeship requirements.[75] It attempted to direct investments to low-income and so-called energy communities through enhanced tax credits for investments in certain locales.[76] And it removed several barriers that long kept publicly owned utilities and energy cooperatives from being able to invest in renewable energy the same way that large corporations can.[77] If the law had survived current political vicissitudes, it would have channeled at least some clean energy investments into tangible redistributions of wealth and power. Jerry may have been able to afford solar-powered electricity from a community-owned solar array sponsored by his rural electric cooperative to replace his coal stove, with no direct limits on consumption necessary.

Should a bill like the IRA be considered a "non-reformist reform"—one that changes the power dynamics of climate lawmaking and not just the investment returns available on clean energy technologies? I agree with Akbar that it is for movements, not academics, to deliver the final verdict here.[78] For my own part, I wish the IRA had done so much more in this regard. But these anti-climate-fundamentalist components of the Act were hard fought and scarcely won, via what journalist Kate Aronoff describes as "an extraordinarily fragile political coalition."[79] It is commendable that even within this fragility, committed activists and policymakers eked out some situated freedom—that is, some possibility of economic and social change toward a political order less beholden to fossil fuel interests.[80] And this, I would argue, is policymakers' ethical duty: to craft policies that not only promote new technologies but also shift the political terrain toward the potential for more radical climate responsivity. That probably does not translate to more conversations about population or consumption *as such*. It does translate to resisting purely techno-optimist solutions that do little more than kick the proverbial can of hard climate truths down the road.[81]

That said, I worry that I may have been wrong about the IRA's potential. It may be that passing legislation like the IRA is another form of denial—of the seriousness of the problem, of the scale of the necessary solutions, of the precious little time left to respond. Under Stanczyk's version of climate justice, perhaps it is cowardly of policymakers to advance such legislation, knowing full well that it cannot prevent numerous disasters from befalling even the presently living. Laws can retrench power as well as shift it, and it was not for nothing that Shell USA, BP America, and Ford joined a letter supporting passage of the Act.[82]

That brings me to one of Stanczyk's final posited conundrums—one of the hardest. He asks "how we, you and I, should reason about our moral obligations once we admit to ourselves that the powerful will predictably not obey any of the most important imperatives of climate ethics."[83] I certainly cannot do justice to this question in my closing paragraph, but I want to offer a few modest thoughts springing from my comments. For the many carbon emissions that are structurally constrained, only state action can provide effective intervention. Our ethical duties therefore cannot end with a focus on our own consumption and population-related choices—they must extend to engagement in political change. If, in the face of collective action and the push for non-reformist reforms, elites and policymakers continue to "respond morally wrongly to the burgeoning human fallout" of climate change,[84] then the next frontier of hard moral questions is how far we should each be willing to go in resistance.

Notes

1 See Lucas Stanczyk, "On the Moral Challenge of the Climate Crisis," in this volume.

2 Stanczyk, "On the Moral Challenge of the Climate Crisis."

3 See Clint Wallace and Shelley Welton, "Taxing Luxury Emissions," *Cornell Law Review* 109, no. 5 (2024): 1153–1231.

4 See Stanczyk, "On the Moral Challenge of the Climate Crisis," in this volume.

5 As Stanczyk traces, the "wedge" concept of climate policy—though not these wedges in particular—stems from Stephen Pacala and Robert Socolow, "Stabilization Wedges: Solving the Climate Prob-

lem for the Next 50 Years with Current Technologies," *Science* 305, no. 5686 (2004): 968–972.

6 Stancyzk, "On the Moral Challenge of the Climate Crisis," in this volume.

7 Inflation Reduction Act, H.R. 5376, 117th Congress, signed August 16, 2022.

8 Otto von Bismarck, St. Petersburgische Zeitung, August 11, 1867.

9 Stanczyk, "On the Moral Challenge of the Climate Crisis," in this volume.

10 Stanczyk, "On the Moral Challenge of the Climate Crisis."

11 Felix Creutzig, Joyashree Roy et al., "Demand, Services, and Social Aspects of Mitigation," in *Climate Change 2022: Mitigation of Climate Change. Contribution of Working Group III to the Sixth Assessment Report of the Intergovernmental Panel on Climate Change*, ed. R. Shukla, J. Skea, R. Slade, A. Al Khourdajie et al. (Intergovernmental Panel on Climate Change, April 2022).

12 Lucas Chancel, "Global Carbon Inequality over 1990–2019," *Nature Sustainability* 5 (2022): 931–938, at 935; Oxfam, "Confronting Carbon Inequality," September 21, 2020, 6.

13 Lucas Chancel, Thomas Piketty, Emmanuel Saez, and Gabriel Zucman, "World Inequality Report 2022" (World Inequality Lab, 2022), 130, available at https://wir2022.wid.world.

14 See Chancel, "Global Carbon Inequality over 1990–2019"; see also Douglas A. Kysar, "Ways Not to Think About Climate Change," in this volume; Wallace and Welton, "Taxing Luxury Emissions."

15 Stanczyk, "On the Moral Challenge of the Climate Crisis," in this volume.

16 Lukas Tank, "Against the Budget View in Climate Ethics," *Critical Review of International Social and Political Philosophy* (2022), at 3 (internal citations omitted).

17 We did not inquire about meat consumption for our project, so I cannot report on this aspect of Jerry's carbon emissions.

18 See Henry Shue, "Subsistence Protection and Mitigation Ambition: Necessities, Economic and Climatic," *British Journal of Politics and International Relations* 21, no. 2 (2019): 251–262.

19 See Tim Hayward, "Human Rights Versus Emissions Rights: Climate Justice and the Equitable Distribution of Ecological Space," *Ethics and International Affairs* 21, no. 4 (2007): 431–450, at 441 ("As long as people are locked into a carbon-dependent economic system they have a right not to be deprived of their basic subsistence rights in virtue of that fact.").

20 See Jared Starr, Craig Nicolson, Michael Ash, Ezra M. Markowitz, and Daniel Moran, "Assessing U.S. Consumers' Carbon Footprints Reveals Outsized Impact of the Top 1%," *Ecological Economics* 205 (2023): 1–27, at 10 (noting that "[h]igh-income households have significant agency [and] discretionary spending").

21 See, e.g., Bill Gates, *How to Avoid a Climate Disaster: The Solutions We Have and the Breakthroughs We Need* (New York: Vintage Books, 2021); Michael Bloomberg and Carl Pope, *Climate of Hope: How Cities, Businesses, and Citizens Can Save the Planet* (New York: St. Martin's Press, 2017); see also William D. Nordhaus, *The Climate Casino* (New Haven, CT: Yale University Press, 2013), 188 (adopting a "descriptive" approach to discounting to "reflect[] the reality that capital is scarce, that societies have valuable alternative investments, and that climate investments should compete with investments in other areas"); Nils Petter Gleditsch, "This Time Is Different! Or Is It? NeoMalthusians and Environmental Optimists in the Age of Climate Change," *Journal of Peace Research* 58, no. 1 (2021): 177–185 (tracing some key tenets of "cornucopian" authors writing against neo-Malthusians). On climate techno-optimism, see Sofia Ribeiro and Viriato Soromenho-Marques, "The Techno-Optimists of Climate Change: Science Communication or Technowashing?," *Societies* 12, no. 2 (2022): 64.

22 A recent analysis of IPCC modeling by the organization Carbon Brief concluded: "All pathways that limit warming to 1.5C or 2C involve substantial levels of CDR [carbon dioxide removal] between 2020 and 2100, ranging from 450 to 1,100 GtCO2." Steve Smith, Jan Minx, Greg Nemet, and Oliver Geden, "Guest Post: The State of 'Carbon Dioxide Removal' in Seven Charts," *CarbonBrief*, January 19, 2023, www.carbonbrief.org.

23 See, e.g., Jacopo Prisco, "This Solar-Powered Plane Could Stay in the Air for Months," *CNN*, May 5, 2022; Joanna Thompson, "Lab-Grown Meat Approved for Sale: What You Need to Know," *Scientific American*, June 20, 2023.

24 See Thomas Malthus, "An Essay on the Principle of Population" (1798); Paul Ehrlich and Ann Ehrlich, *The Population Bomb* (New York: Sierra Club-Ballantine Books, 1968); Donella Meadows, Dennis Meadows, Jørgen Randers, and William Behrens, *The Limits to Growth: A Report for the Club of Rome's Project on the Predicament of Mankind* (New York: Universe Books, 1972). On the history of these doomsayers, see Paul Sabin, *The Bet: Paul Ehrlich, Julian Simon, and Our Gamble over Earth's Future* (New Haven, CT: Yale University Press, 2013) (chronicling the infamous bet between cornucopian Simon and doomsayer Ehrlich); David Deming, "M. King Hubbert and the Rise and Fall of Peak Oil Theory," *AAPG Bulletin* 107, no. 6 (2023): 851–861.

25 Whether corporations that putatively plan for carbon capture and storage to absolve their climate sins are doing enough to bring about its commercialization is an interesting question that might tie into Stanczyk's theory about what the present generation owes in its climate planning. See Carlos Anchondo, Jason Plautz, and Zach Bright, "EPA Says Carbon Capture is Within Reach. Utilities Aren't Biting," *E&E News*, July 11, 2023, www.eenews.net.

26 See "Implementation and Action Toward Net Zero," Southern Company (September 2020): 3, available at www.southerncompany.com.

27 See Amanda C. Borth and Simon Nicholson, "A Deliberative Orientation to Governing Carbon Dioxide Removal: Actionable Recommendations for National-Level Action," *Frontiers in Climate* 3 (July 2021): 684209; Hiroko Tabuchi, "Switzerland Is Paying Poorer Nations to Cut Emissions on Its Behalf," *New York Times*, November 22, 2022.

28 See, e.g., Douglas A. Kysar, *Regulating from Nowhere: Environmental Law and the Search for Objectivity* (New Haven, CT: Yale University Press, 2010); John S. Applegate, "Embracing a Precautionary Approach to Climate Change, in Economic Thought and U.S. Climate Change Policy," in *Economic Thought and U.S. Climate Change Policy*, ed. David M. Driesen (Cambridge, MA: MIT Press, 2010); Noah M. Sachs, "Rescuing the Strong Precautionary Principle from Its Critics," *University of Illinois Law Review* 2011, no. 4 (2011): 1285–1338.

29 Stanczyk, "On the Moral Challenge of the Climate Crisis," in this volume.

30 See, e.g., Colin Hickey, Travis N. Rieder, and Jake Earl, "Population Engineering and the Fight against Climate Change," *Social Theory and Practice* 42, no. 4 (2016): 845–870 (charting a spectrum of potential interventions to influence population and condoning only some of them); Eileen Christ, William J. Ripple, Paul R. Ehrlich, William E. Rees, and Christopher Wolf, "Scientists' Warning on Population," *Science of the Total Environment* 845 (2022): 157166. Donna Haraway's "Anthropocene, Capitalocene, Plantationocene, Chthulucene: Making Kin," *Environmental Humanities* 6, no. 1 (2015): 159–165, is one of the more surprising forays into this topic, in which she provocatively explains: "I know 'population' is a state-making category, the sort of 'abstraction' and 'discourse' that remake reality for everybody, but not for everybody's benefit. But blaming Capitalism, Imperialism, Neoliberalism, Modernization, or some other 'not us' for ongoing destruction webbed with human numbers will not work either." Quote on p. 164 n.7; see also *Making Kin not Population*, ed. Adele Clarke and Donna Haraway (Chicago: University of Chicago Press, 2018).

31 See, e.g., Jade S. Sasser, *On Infertile Ground: Population Control and Women's Rights in the Era of Climate Change* (New York: New York Univer-

sity Press 2018); Nayantara Sheoran Appleton and Danya Labau, "Critical Engagements on *Making Kin Not Population*: An Epistolary Review Essay," *American Anthropologist* 124, no. 4 (2020): 891–899, at 892 ("[T]he focus on population size singularly allows the elite to virtue signal by not having children while continuing with measurably more-damaging lifestyles."); Anne Hendrixson, Diana Ojeda, Jade S. Sasser, Sarojini Nadimpally, Ellen E. Foley, and Rajani Bhatia, "Confronting Populationism: Feminist Challenges to Population Control in an Era of Climate Change," *Gender, Place, & Culture* 27, no. 3 (2020): 307–315.

32 Stanczyk, "On the Moral Challenge of the Climate Crisis," in this volume.

33 See Food and Agricultural Association of the United Nations, *Global Food Losses and Food Waste: Extent, Causes, and Prevention* (Rome: 2011).

34 See Stanczyk, "On the Moral Challenge of the Climate Crisis," in this volume.

35 See Chancel, "Global Carbon Inequality over 1990–2019"; Kysar, "Ways Not to Think About Climate Change," in this volume.

36 *Our World in Data*, "CO2 Emissions Per Capita," https://ourworldindata.org. See also Hickey et al., "Population Engineering and the Fight against Climate Change," at 855–56 ("Although it would be difficult to lower the fertility rate in the United States from 1.9 to, say, 1.4, such a reduction would have a massive impact on both near-term and long-term global GHG emissions—much more even than proportionally larger fertility reductions in sub-Saharan Africa."). I'm using the US as a stand-in for over-consumption, but of course, there is extreme inequality leading to vast gulfs in consumption patterns in every country. The US just happens to be particularly high-consuming both on an average per-capita basis and in terms of its super-emitters, making it an easy target. See Chancel, "Global Carbon Inequality over 1990–2019."

37 See Susanne Shultz, "The Neo-Malthusian Reflex in Climate Politics: Technocratic, Right Wing and Feminist References," *Australian Feminist Studies* 36, no. 110 (2021): 485–502, at 486, 492 (strongly condemning the "neo-Malthusian reflex" in climate change policy and arguing against the "demographisation" of the climate-population nexus, by which she means the "re/interpretation of social crises as demographic ones"); Dorothy Roberts, "Reproductive Justice, Not Just Rights," *Dissent* 62, no. 4 (2015): 79–82, at 79 (tracing the history of the reproductive justice movement as developed by Black feminists to include "not only a woman's right not to have a child, but also the right to have children and to raise them with dignity in safe, healthy, and supportive environments"); Appleton and Labau, "Critical Engagements on *Making Kin Not Population*: An Epistolary Review Essay," at 893 (warning against using "population" "as a concept, unmoored from

history and politics, rather than as a historically contingent and murderous invention of the economic, social, and biological sciences").

38 See, e.g., Christ et al., "Scientists' Warning on Population."

39 See, e.g., R. John Aitken, "The Changing Tide of Human Fertility," *Human Reproduction* 37, no. 4 (2022): 629–638, at 629 ("The fundamental cause of human fertility decline is prosperity"); Emmanuel Obi, "Effects of High Fertility on Economic Development," American Economic Association, Working Paper (2019), at 2 ("[A] broad consensus has developed over time that as incomes rise, fertility tends to fall. There is little debate about the causal relationship between rising prosperity and declining fertility. Generally speaking, there has been a uniformly high correlation between national income growth and falling birth rates, and between family incomes and fertility.").

40 See, e.g., "Global Fertility Has Collapsed, with Profound Economic Consequences," *The Economist*, June 1, 2023, www.economist.com.

41 See Mark Budolfson and Dean Spears, "Population Ethics and the Prospects for Fertility Policy as Climate Mitigation Policy," *Journal of Development Studies* 57, no. 9 (2021): 1499–1510, at 1500 ("[T]he region where fertility rates are high enough that there is scope, in principle, for faster decline is sub-Saharan Africa—where emissions per capita are currently low.").

42 See Obi, "Effects of High Fertility on Economic Development." To be extremely clear, this is not an argument against pursuing policies that empower people everywhere to manage choices surrounding reproduction and parenthood as part of a more capacious commitment to reproductive justice—but I'm not sure carbon emissions should drive the strategies pursued. Cf. Hickey, Rieder, and Earl, "Population Engineering and the Fight against Climate Change," at 854 (advocating that "choice-enhancing interventions" "are not only permissible, but obligatory, as they are means of ensuring equal access to basic goods").

43 See Budolfson and Spears, "Population Ethics and the Prospects for Fertility Policy as Climate Mitigation Policy," 1502 (weighing these complexities and arguing that fertility reduction is not "likely to be a quantitatively successful climate mitigation policy").

44 See Dobbs v. Jackson Women's Health Org., 597 U.S. 215 (2022).

45 See, e.g., Jake Earl, Colin Hickey, and Travis N. Rieder, "Fertility, Immigration, and the Fight Against Climate Change," *Bioethics* 31, no. 8 (2017): 582–589, at 583.

46 See Earl, Hickey, and Rieder, "Fertility, Immigration, and the Fight Against Climate Change"; see also Jamie Draper, "The Environmental Argument for Immigration Restrictions: A Critique," in this volume.

47 Stanczyk, "On the Moral Challenge of the Climate Crisis," in this volume.
48 Stanczyk, "On the Moral Challenge of the Climate Crisis," 16 n.32.
49 Stanczyk, "On the Moral Challenge of the Climate Crisis," 7–8.
50 See Stanczyk, "On the Moral Challenge of the Climate Crisis," 9–10.
51 Stanczyk, "On the Moral Challenge of the Climate Crisis," 10.
52 Stanczyk, "On the Moral Challenge of the Climate Crisis," 11.
53 Cf. Kysar, *Regulating from Nowhere*, 177 (suggesting that one promising starting point for moving beyond the Non-Identity Problem might be "conceiving of future generations as coherent collective entities, rather than merely as individual lives-in-waiting").
54 See, e.g., Winona LaDuke, *All Our Relations: Native Struggles for Land and Life* (Chicago: Haymarket, 2016), chap. 10, "The Seventh Generation."
55 See Stanczyk, "On the Moral Challenge of the Climate Crisis," in this volume.
56 Stanczyk, "On the Moral Challenge of the Climate Crisis."
57 Stanczyk, "On the Moral Challenge of the Climate Crisis."
58 Matthew T. Huber, *Climate Change as Class War: Building Socialism on a Warming Planet* (New York: Verso, 2022), 147–151.
59 See Abha Bhattarai, "Infrastructure and Green Energy Spending Are Powering the Economy," *Washington Post*, July 28, 2023.
60 See John Bistline et al., "Emissions and Energy Impacts of the Inflation Reduction Act," *Science* 380, no. 6652 (2023): 1324–1327.
61 Stanczyk, "On the Moral Challenge of the Climate Crisis," in this volume; see also Naomi Klein, *The Shock Doctrine: The Rise of Disaster Capitalism* (New York: Henry Holt, 2007).
62 Stanczyk, "On the Moral Challenge of the Climate Crisis," in this volume ("[T]he worst effects of climate change will come not from the storms or the fires or the simultaneous crop failures. The worst effects will come when the economic elites and other people respond morally wrongly to the burgeoning human fallout.").
63 See Jedediah Britton-Purdy, David Singh Grewal, Amy Kapcyznski, and K. Sabeel Rahman, "Building a Law-and-Political-Economy Framework: Beyond the Twentieth-Century Synthesis," *Yale Law Journal* 129, no. 6 (2020): 1784–1835.
64 See, e.g., Britton-Purdy et al., "Building and Law-and-Political-Economy Framework," 1820 (calling for attention to "the constitutive power of law to create endowments that shape all voluntary bargains, the market power that legal structures enable, and the political power that may arise from differential endowments, market power, or ways that legal rules

insulate economic power from democratic reordering"); Amna A. Akbar, "Non-Reformist Reforms and Struggles over Life, Death, and Democracy," *Yale Law Journal* 132, no. 8 (2023): 2497–2577, at 2508; Robert W. Gordon, "Critical Legal Histories," *Stanford Law Review* 36, (1984): 57–125, at 118 (arguing that "[w]hat [legal] structures 'determine' is not any particular set of social consequences but the categories of thought and discourse wherein political conflict will be carried out"). There is a related though narrower movement in political science focused on "policy feedback" that is also relevant, inasmuch as it explores how policies create their own political economies that in turn create new landscapes of political opportunity. See, e.g., Leah Stokes, *Short Circuiting Policy: Interest Groups and the Battle over Clean Energy and Climate Policy in the American States* (Oxford: Oxford University Press, 2020); Eric Biber, "Cultivating a Green Political Landscape: Lessons for Climate Change Policy from the Defeat of California's Proposition 23," *Vanderbilt Law Review* 66, no. 2 (2013): 399–462.

65 See Samuel Moyn, "Reconstructing Critical Legal Studies," *Yale Law Journal* 134, no. 1 (2024): 77–122; Samuel Moyn, "From Situated Freedom to Plausible Worlds," in *Contingency in International Law: On the Possibility of Different Legal Histories*, ed. Ingo Venztke and Kevin Jon Heller (Oxford: Oxford University Press, 2021).

66 Akbar, "Non-Reformist Reforms and Struggles over Life, Death, and Democracy," at 2507.

67 Akbar, "Non-Reformist Reforms and Struggles over Life, Death, and Democracy," at 2507.

68 Shalanda H. Baker, "Anti-Resilience: A Roadmap for Transformational Justice within the Energy System," *Harvard Civil Rights-Civil Liberties Law Review* 54 (2019): 1–48, at 15 n.64.

69 See Baker, "Anti-Resilience: A Roadmap for Transformational Justice within the Energy System," 15–18.

70 See, e.g., Maxine Burkett, "Root and Branch: Climate Catastrophe, Racial Crises, and the History and Future of Climate Justice," *Harvard Law Review Forum* 134, no. 6 (2021): 326–339; Andreas Malm, *Fossil Capital* (New York: Verso, 2016); Timothy Mitchell, *Carbon Democracy: Political Power in the Age of Oil* (New York: Verso, 2011); Amitov Ghosh, *The Nutmeg's Curse: Parables for a Planet in Crisis* (Chicago: University of Chicago Press, 2021).

71 See Daniel Aldana Cohen, Alyssa Battistoni, Thea Riofrancos, and Kate Aronoff, *A Planet to Win: Why We Need a Green New Deal* (New York: Verso, 2019), 18, 89–92, 187–191; Rhiana Gunn-Wright and Robert Hockett, "Mobilizing for a Just, Prosperous, and Sustainable Economy: The Green New Deal," Cornell Law School Research Paper No. 19-09 (2019), 6.

72 See, e.g., H.R. Res. 109, 116th Cong. (1st Sess. 2019) (House Resolution introduced by Senator Markey and Representative Ocasio-Cortez calling for community ownership of resources, a federal jobs guarantee, and "high-quality health care" and "affordable, safe, and adequate housing" for all Americans).

73 Cf. Alyssa Battistoni, "Picking Winners," *Sidecar*, November 24, 2021, https://newleftreview.com (describing the IRA as "more Silicon Valley than the Tennessee Valley Authority" and "tak[ing] cues not from the public investment-driven Green New Deal of Alexandria Ocasio-Cortez but from the innovation-oriented Green New Deal of the late 2000s").

74 McKinsey estimates that of the approximately $394 million predicted by congressional estimates to be dispersed in IRA tax credits, $216 billion will go to corporations. "The Inflation Reduction Act: Here's What's In It," McKinsey & Company, October 24, 2022, www.mckinsey.com. See also Brett Christophers, "Why Are We Allowing the Private Sector to Take Over Our Public Works?," *New York Times*, May 8, 2023; Daniela Gabor, "The (European) Derisking State," *Stato e Mercato* 127, no. 1 (2023): 53–84.

75 See Hannah Sachs and David Foster, "Job Quality: The Keystone of Clean Energy Industrial Policy," Energy Futures Initiative, August 8, 2023, www.efifoundation.org.

76 Sachs and Foster, "Job Quality: The Keystone of Clean Energy Industrial Policy."

77 See Chirag Lang, "Direct Pay: An Uncapped Promise of the Inflation Reduction Act," Center for Public Enterprise, March 30, 2023, www.publicenterprise.org.

78 Akbar, "Non-Reformist Reforms and Struggles over Life, Death, and Democracy," at 2535–2536.

79 Kate Aronoff, "The Case for Pool Party Progressivism," *New Republic*, August 15, 2023, www.newrepublic.com.

80 See Moyn, "Reconstructing Critical Legal Studies," *Yale Law Journal* 134, no. 1 (2024): 77–122. ("Legal orders produce agency sufficient to change them . . . though they differ radically in the extent to which they do so.")

81 See Stanczyk, "On the Moral Challenge of the Climate Crisis," in this volume.

82 See "Business Support Statement for the Inflation Reduction Act" (n.d.), www.ceres.org.

83 Stanczyk, "On the Moral Challenge of the Climate Crisis," in this volume.

84 Stanczyk, "On the Moral Challenge of the Climate Crisis."

10

THE ENVIRONMENTAL ARGUMENT FOR IMMIGRATION RESTRICTIONS

A CRITIQUE

JAMIE DRAPER

Egalitarians have good reasons to care about both climate change and migration. The impacts of climate change tend to fall hardest on those who are already among the worst-off, exacerbating inequality and entrenching poverty.[1] Immigration restrictions in high-income states lock the worst-off out of opportunities to increase their income, wealth, and wellbeing.[2] Those concerned by global inequality thus tend to advocate both aggressive climate mitigation policies and less restrictive immigration policies in high-income states.[3]

But what if these two goals conflict? One possible concern about migration from low- to high-income states is that it increases the population of high-emitting states, thereby increasing greenhouse gas emissions and hindering the project of climate mitigation. If this is correct, it suggests that there may be a conflict between the egalitarian goals of loosening immigration restrictions in high-income states and taking aggressive action to mitigate climate change. What should egalitarians make of this apparent conflict?

The *environmental argument for immigration restrictions* suggests that high-income states ought to restrict immigration in order to promote environmental goals such as mitigating climate change.[4] In this chapter, I critically assess the environmental argument for immigration restrictions. First, I distinguish between several different versions of the environmental argument and reconstruct the most plausible version of it. Next, I clarify the empirical basis of the environmental argument. Then, I examine three counterarguments to the environmental argument: first, that migration-related emissions ought to be treated as special when it comes to climate mitigation; second, that it is inconsistent to promote human development abroad and restrictive immigration policies at home; and third, that high-emitting states cannot appeal to environmental arguments as a justification for exclusion. I argue that the first counterargument fails, that the second is partially successful, and that the third counterargument is successful. The upshot is that egalitarians concerned about climate change need not advocate for restrictive immigration policies in high-income states.

Before making these arguments, it is worth presenting three clarificatory remarks. First is that my focus in this chapter is on the consequences of migration for climate change, not the consequences of climate change for migration. The impacts of climate change interact with patterns of migration and displacement in empirically and morally complex ways, which raise important questions of justice that I have explored in greater detail elsewhere.[5] Here, I focus instead on the impacts of migration from low- to high-income states on climate mitigation.

Second, in this chapter I attempt to engage with the environmental argument in its most plausible form. In political debates, the environmental argument is often made in bad faith by those who cloak their nativism in an environmentalist garb. Across Europe, political parties on the far right are increasingly attempting to weaponize environmental issues to bolster anti-immigrant sentiment.[6] And even within mainstream environmental movements, there have always been those whose environmentalism is intertwined with anti-immigrant hostility.[7] But the environmental argument for immigration restrictions is also made in good faith by those who believe there is a genuine tradeoff between openness to immigration and environmental protection. It is worth engaging with this argument at its best, to show why those genuinely committed to tackling climate change need not endorse restrictive immigration policies.

Third, I engage here with the environmental argument as it is directed toward egalitarians. Although it presents a challenge to anyone who favors more open immigration policies, the environmental argument—or at least the version of it that I focus on—presents an especially important challenge for egalitarians because it appeals to commitments that they already share. Egalitarians who seek to improve the position of the globally worst-off argue that this goal provides a good reason to loosen immigration restrictions in high-income states. But if the environmental argument is successful, then it shows that this commitment to global equality actually speaks *against*, rather than in favor of, loosening immigration restrictions in high-income states.

The Environmental Argument for Immigration Restrictions

The most explicit defense of the environmental argument for immigration restrictions is made by Philip Cafaro and Winthrop Staples III.[8] Cafaro and Staples focus on restricting immigration to the United States, but I reconstruct their argument in more general (and simplified) terms here. As we will see, there are several versions of the environmental argument. But the argument has a common core, which can be set out as follows:

> P1. Immigration is a significant cause of population growth in high-income states.
> P2. Population growth in high-income states is a significant cause of environmental degradation.
> P3. We have a moral duty to reduce environmental degradation.
> C. Therefore, we have a moral duty to reduce immigration to high-income states.[9]

At first glance, this argument appears to provide reasonable grounds for immigration restrictions. The first two premises are at least facially plausible empirical claims, which I examine more closely in the next section. The third premise is a normative premise, which appeals to a moral duty that we can expect to be accepted by egalitarians. Cafaro and Staples suggest that the argument rests only "on a straightforward commitment to mainstream environmentalism, easily confirmed empirical premises, and logic."[10]

Before evaluating this argument, it is worth examining P3, its key normative premise, in greater detail. A "moral duty to reduce environmental degradation" could be understood in multiple ways. Clarifying this duty reveals that there are several different versions of the environmental argument for immigration restrictions and allows us to identify the version of the argument that presents the most significant challenge for egalitarians.

One distinction between different versions of the environmental argument concerns the *scope* of P3. The duty to reduce environmental degradation could apply specifically to environmental degradation within the receiving society, or it could apply to environmental degradation throughout the world. We can call the former version of the argument the *local* argument for immigration restrictions, and the latter the *global* argument for immigration restrictions. In one part of their discussion, Cafaro and Staples make the local argument for immigration restrictions: They argue that the US government should restrict immigration because immigration increases urban sprawl and consequent habitat destruction within the US.[11] In another part of their discussion, they make the global argument for immigration restrictions, focusing on the role that immigration-related population growth plays in the US's contributions to global climate change.[12] David Miller also makes a version of the global argument.[13] He suggests that high-income states ought to restrict immigration as part of their efforts to tackle climate change, because "immigrants who adopt Western lifestyles will consume more and induce more carbon emissions through their demand for energy" such that "migration is likely to be bad news for the planet overall."[14]

Another distinction concerns the different possible *grounds* for P3, the duty to reduce environmental degradation. Why do we have a duty to reduce environmental degradation? One answer to this question is because of the negative (and unevenly distributed) consequences of environmental degradation for other *humans*, including future generations of humans. On this view, the duty to reduce environmental degradation is continuous with other duties of distributive justice, in that it concerns the fair distribution of (environmental) benefits and burdens within human communities. Another answer to this question is that we have a duty to reduce environmental degradation because of the negative consequences of environmental degradation for *non-human* entities such as non-human animals, species, or ecosystems. On this view, our duty to reduce environmental degradation is part of a broader ecological ethic that treats some non-human entities as bearers of intrinsic value or subjects of moral concern. We can call the former version of the argument the *environmental* argument for immigration restrictions and the latter version the *ecological* argument for immigration restrictions.[15] Cafaro and Staples say relatively little about the grounds of their argument, except that they take the duty to reduce environmental degradation to follow from a commitment to mainstream environmentalism. But when they do reflect on this commitment, they refer most clearly to the ecological argument, arguing that environmentalism involves a commitment to "generous sustainability" which includes the goal of "sharing the landscape generously with non-human beings."[16] They do, however, suggest that their argument can also be endorsed by those who hold anthropocentric views, on the basis that migration has negative environmental consequences for "future generations in the United States and abroad."[17] And in other work defending immigration restrictions, Cafaro has based his argument on the claim that "justice demands vigorous action from this current generation of Americans to prevent potentially catastrophic climate change . . . for ourselves and our children, for the sake of the world's poorest people, for future generations."[18]

These possible interpretations of the P3 allow us to distinguish between four versions of the environmental argument for immigration restrictions: the *local-ecological* argument, the *global-ecological* argument, the *local-environmental* argument, and the *global-environmental* argument. The local-ecological argument concerns the environmental impact of immigration on non-human entities within the receiving society. The global-ecological argument concerns the environmental impact of immigration on non-human entities throughout the world. The local-environmental argument concerns the environmental impact of immigration on the interests of humans within the receiving society. And the global-environmental argument concerns the environmental impact of immigration on the interests of humans globally.

I take the global-environmental argument to be the strongest version of the environmental argument, so it will be the primary target of my critique. The global-environmental argument presents a particular challenge to egalitarians, since it says that immigration restrictions in high-income states are justified because such restrictions are an important way of protecting the interests of those who are most vulnerable to climate impacts. And those who are most vulnerable to climate impacts are, by and large, the globally worst-off. The local and ecological versions of the argument, by contrast, are less plausible candidates for a successful environmental argument for immigration restrictions.

The local argument evaluates immigration policies only in terms of their impacts on the receiving society. But why should we restrict our focus in this way? Advocates of the local argument need to claim that—at least when it comes to environmental degradation—the interests of those (humans or non-human entities) within the receiving society should be prioritized over those abroad. For their part, Cafaro and Staples appear to endorse the local argument, in both its ecological and environmental variants. They "plead guilty to a special concern for America's wildlife and wetlands" and suggest that "environmentalism necessarily involves love, connection and efforts to protect particular places."[19] They also argue that an additional (though non-environmental) reason to restrict immigration is that it has negative economic consequences for the worst-off within the receiving society.[20] But claims like these are likely to be controversial, especially for those who would otherwise favor more open immigration policies because of their effects on global inequality. The assumption that the (human or non-human) inhabitants of the receiving society should have a moral priority imposes a significant justificatory burden on the local argument and makes it less promising as an argument for immigration restrictions that is directed at egalitarians.

The ecological argument presupposes a non-anthropocentric ethic, according to which non-human entities such as non-human animals, species, and ecosystems have moral status or intrinsic value. Cafaro and Staples suggest that some such non-anthropocentric ethic "captures the core of environmentalism," though they leave the precise content of this ethic unspecified.[21] A more minimal version of this ethic might attribute moral status to some non-human animals, while a broader version would attribute moral status or intrinsic value to ecological wholes such as species and ecosystems. Different versions of this ethic will be more and less controversial, and some egalitarians will accept that the interests of non-human entities should count when it comes to global justice.[22] But the claim involved in the ecological argument is not only that the interests of non-human entities *count*, but moreover that those interests can override important human interests, such as the interests that would be served through migration policies that improve the position of the globally worst-off.[23] Like the local argument, the ecological argument thus also requires us to accept claims that are likely to be controversial among those who would typically favor more open immigration policies because of the role that they can play in improving the position of the worst-off.

The global-environmental version of the environmental argument for immigration restrictions avoids both of these sets of controversial assumptions. It does not assume any priority for local (human or non-human) victims of environmental degradation, nor does it assume any controversial theory of moral status or intrinsic value that suggests that the interests of non-human entities can outweigh the interests of humans. It is entirely consistent with an anthropocentric, cosmopolitan egalitarianism. This makes it a particularly promising argument in the context of debates with those who otherwise endorse open immigration policies as a way of reducing global inequalities.

Migration, Population Growth, and Climate Change

The environmental argument for immigration restrictions claims that migration from low- to high-income states, because it leads to population growth in the receiving society, has negative environmental consequences. In the global-environmental argument, the relevant negative consequences are increased greenhouse gas emissions, which aggravate the threat of climate change to important human interests. None of the arguments that I consider in this chapter reject this empirical claim, but it is nonetheless important to clarify the relationship between migration, population growth, and climate change within the environmental argument.

One classic formula for determining environmental impact is the IPAT formula, which treats environmental impact (I) as a function of population (P), affluence (or consumption) (A), and technological development (T).[24] The IPAT formula offers an intuitive explanation of the role of population growth in environmental degradation: If population levels increase while levels of affluence and technological development remain the same, then we can expect overall environmental impact to increase. Population acts as a multiplier for the environmental impact of consumption at a given level of technological development. The IPAT equation is certainly not perfect. For one thing, its variables are not independent of each other. The pace of technological development, for example, is itself affected by the dynamics of economic growth.[25] For another, its variables may not affect environmental impact in the way that the formula predicts. For example, technological development might induce increases in consumption that offset efficiency savings, or increases in affluence above a threshold may reduce rather than increase environmental impact.[26] Still, despite these issues, the IPAT equation is a useful heuristic, and helps us to get a better picture of the roles that migration and population growth may play in climate change.

Migration itself does not create more people, and so it does not affect population growth directly. It may have an *indirect* effect on population growth, if migration changes fertility behavior such that migrants (or their children) have more or fewer children than they would otherwise have. But when it comes to the fertility behavior of immigrants, the evidence suggests that fertility rates tend to decline among migrants who move to high-income states and their children, converging over time with lower fertility rates among non-immigrants.[27] If anything, this suggests that—other things being equal—we should expect migration from low- to high income to *reduce* rather than increase global population, although demographers typically stress that other factors such as women's educational attainment and access to contraception are much more important than migration.[28]

When we talk about population growth resulting from immigration, then, we are not talking about *absolute* population growth at the global level. Rather, we are talking about *relative* population growth within the receiving society. The central problem with migration from low- to high-income states, according to the environmental argument, is that it increases the size of the population in places where greenhouse gas emissions are already high.

The inhabitants of high-income states consume environmental resources at an unsustainable rate, by emitting more than their fair share of the greenhouse gases than can be safely absorbed into the Earth's atmosphere without leading to dangerous climate impacts. Migration from low- to high-income states increases their share of the global population, thus increasing overall emissions and aggravating the problem of climate change. Restrictive immigration policies in high-income states, so the argument goes, would reduce the number of people who are emitting too much, by limiting population growth in high-income states.

The empirical relationship that is ultimately important from the point of view of the environmental argument is thus the relationship between migration from low- to high-income states and overall greenhouse gas emissions. There is relatively little work that focuses directly on this relationship, but one recent analysis does provide suggestive evidence of a link, finding that the CO_2 emissions attributable to international migrants increased by 65% in the period from 1995 to 2015, while global emissions rose by roughly 33% in the same period.[29] The authors of the study hypothesize that the best explanation for this is that those who migrate from poorer to richer states tend to change their lifestyles, and their consumption of goods such as housing, infrastructure, and health care will tend to have a greater CO_2 intensity.[30] This evidence is certainly not conclusive, but for the purposes of the arguments that I develop here, I will accept the empirical claim that migration from low- to high-income states will tend to increase greenhouse gas emissions overall.

There are, however, two qualifications that are worth adding to this empirical claim. The first is that there is some evidence to suggest that immigrants tend to have a lower environmental impact than non-immigrants, with studies finding lower rates of CO_2 and N_2O emissions or air pollution in areas with a higher share of immigrants (though it is not entirely clear if this is because immigrants make different consumption choices or simply because they tend to be worse off than non-immigrants).[31] These findings suggest that population growth in high-income states due to migration may have a weaker effect on greenhouse gas emissions than equivalent "natural" population growth. If they are correct, then their implication is that the environmental consequences of migration are likely to be less pronounced than the environmental argument suggests.

But this qualification does not undermine the empirical basis of the environmental argument. Even if immigrants emit less than non-immigrants, they may still emit more than they would have otherwise emitted at home, so migration from low- to high-income states may still increase greenhouse gas emissions overall.

The second qualification is that the impact of migration on climate *mitigation* is only part of the story, since migration from low- to high-income states may also play a role in *adaptation* to climate change. When it comes to tackling climate change, another strategy besides reducing climate impacts through mitigation is making those impacts less harmful through adaptation. Elsewhere, I have argued that international migration can, at least in some circumstances, function as an effective and just strategy of adaptation to climate change, because of the role that remittances can play in promoting climate resilience amongst those who remain behind.[32] Recent empirical work by Hélène Benveniste, Michael Oppenheimer, and Marc Fleurbaey also suggests that restrictive immigration policies can exacerbate vulnerability to climate change by keeping people trapped in places where they face greater exposure to climate impacts.[33] Although I do not pursue this line of argument further here, a full accounting of the effects of migration on climate change would need to take into account the adaptive benefits of migration, as well as any costs that it creates in terms of hindering climate mitigation.

What should we take from this clarification of the empirical basis of the environmental argument? There are two important points to draw out for our purposes. First is that population growth in high-income states only matters, from an environmental point of view, because of high levels of emissions within receiving societies. Immigration from low- to high-income states does not increase population levels overall, and so it would not be an environmental issue were it not for high emissions within receiving societies.

Second is that there are other levers apart from population that we can pull in mitigating climate change. High-income states' contributions to climate change are a consequence not only of their population sizes, but also of their levels of consumption and technological development. These states can also reduce their contributions to climate change by reducing consumption, by increasing technological development, or through some combination of both.

Our view about which (combination of) levers we can or should pull in mitigating climate change will depend on a host of highly contested moral and empirical claims about the population, consumption, and technology. Darrel Moellendorf, for example, argues that we should focus on technology, partly because he is optimistic about technological development and partly because of a moral concern that reducing consumption in high-income states will have knock-on effects that exacerbate global poverty.[34] Henry Shue argues against betting on at least some forms of technological development because doing so involves shifting the burden of risk between present and future generations.[35] Advocates of degrowth who are skeptical about technological development typically stress the importance of reducing consumption.[36] There are deep disagreements here, including not only moral disagreements, but also empirical and theoretical disagreements about whether there are necessary limits to economic growth or technological development (and, if so, where exactly those limits lie).[37]

For their part, Cafaro and Staples argue that *both* population growth and consumption should be reduced within high-income states, though they say relatively little about technological development.[38] It is worth making two points about this position. The first is that population growth is *already* declining in high-income states. In 2021, the US population growth rate was 0.3%, a figure that has been steadily declining since 1992. The US's fertility rate in 2021 was 1.66, well below the estimated fertility replacement rate of roughly 2.1.[39] Like other advanced industrialized economies, the US has undergone a "demographic transition" as birth and mortality rates both decrease over time.[40] With current levels of migration, the US's population is growing slowly, and the growth rate is falling. If migration were restricted, then US population levels would significantly decline overall. So, although they talk about preventing population *growth*, Cafaro and Staples's argument is better understood as the claim that we should significantly *reduce* population levels in high-income states.[41]

The second point is that reducing population levels within high-income states only makes sense as a mitigation strategy if emissions within those states remain higher than can be sustained at a global level. Cafaro and Staples argue that "if high consumption is a problem, population growth must be, too."[42] But by the same token, if relative population growth within high-income societies (rather than absolute population growth at the global level) is a problem, this can only be because emissions within those societies remain too high. If the inhabitants of high-income states were to emit at a globally sustainable rate, then there would be no need to reduce their relative share of the global population in order to tackle climate change.

As we will see, these points are important when it comes to justifying immigration restrictions. With the relationship between migration, population growth, and climate change clarified, we can turn to three counterarguments to the environmental argument for immigration restrictions. The first counterargument suggests that migration-related emissions ought to enjoy a special status within climate mitigation; the second suggests that it is inconsistent to promote immigration restrictions and development abroad; and the third suggests that high-emitting states cannot appeal to environmental reasons to exclude would-be immigrants.

Is Migration Special?

Philosophers working on population and climate change have tended to focus on population growth via procreation, not immigration.[43] But their arguments could be extended to migration. Ingrid Robeyns has recently argued that procreation ought to enjoy a special status compared to other emissions-generating activities.[44] Robeyns argues against what she calls the "moral equivalence thesis"—the claim that procreation is "morally on a par" with consumption—and defends the claim that procreation ought to enjoy a special status in our response to the climate crisis.[45] If Robeyns is right, then we might extend her argument: Perhaps migration *also* ought to enjoy a special status when it comes to tackling the climate crisis.

Robeyns makes two arguments. The first is that we should view "procreative parenting" as a minimal capability that ought to be socially guaranteed in our response to the climate crisis rather than as a mere preference.[46] Robeyns argues that we should first ensure that each person is guaranteed a minimal set of valuable capabilities, and only then distribute any remaining emissions entitlements for mere preferences fairly between persons. If we accept this view, and if we accept that the capability of procreative parenting should fall within this minimal set, then emissions from procreative parenting and emissions from consumption activities such as flying for leisure travel would not be morally equivalent.[47] The capability of procreative parenting (the opportunity to have at least one child, but not necessarily more, on Robeyns's view) would be protected within climate mitigation.

Robeyns's second argument is the "argument from human self-understanding," which says that the opportunity to procreate is part of what it means to be a human.[48] This argument depends on a comprehensive interpretation of what it means to live well as a human, which involves understanding humans as "a species that procreates and has developed many social practices and traditions that give meaning to parenting and kinship."[49] If successful, it suggests that procreation is a matter of seeing ourselves as human, such that being denied the opportunity to procreate (at least once, but not necessarily more than once) amounts to dehumanization. It suggests that in our response to the climate crisis, we have special reason to protect procreation, but not mere consumption, since being denied opportunities for procreation is dehumanizing.[50]

Could these arguments be extended to the case of migration? If so, then this would provide a strong counterargument to the environmental argument for immigration restrictions. It would suggest that rather than focusing on reducing the population of high-income states by restricting immigration, we should instead focus on reducing consumption or increasing technological development. But my suggestion is that whatever their prospects for the case of procreation, neither of these arguments provides the basis for a successful response to the environmental argument for immigration restrictions.

On the first argument, the important question for our purposes is whether migration should be included in the minimal set of valuable capabilities that enjoy a protected status in our response to the climate crisis. Migration can certainly be understood as a capability, and it is often a valuable one.[51] As egalitarians have pointed out, migration is often instrumentally valuable as a means of achieving further goods and opportunities, such as access to jobs or education. Since migration, especially from low- to high-income states, can create significant opportunities, it may even be what Jo Wolff and Avner de Shalit refer to as a "fertile functioning."[52] And beyond its instrumental value, the opportunity to migrate may also be non-instrumentally valuable. Those who undertake what Valeria Ottonelli and Tiziana Torresi call a "migration project" view the opportunity to migrate abroad and acquire new experiences as an important part of their life plan, independently of its instrumental benefits.[53]

But even if migration is a valuable capability, this does not mean that it should be counted among the minimal set of capabilities that ought to be guaranteed as a matter of distributive justice. The instrumental value of migration from low- to high-income states suggests that our primary goal should not be to protect migration *itself* within our climate mitigation efforts, but rather to ensure that the worst-off have access to the opportunities for which migration from low- to high-income states is often instrumental.[54] Access to good jobs and education ought to be protected in our response to the climate crisis, but as we will see in the next section, there may be other ways of providing access to these goods besides migration.

The fact that migration is non-instrumentally valuable for some also does not show that it should be protected as a basic capability. After all, there are many other emissions-generating activities which may be non-instrumentally valuable for their participants that we may not want to treat as special, such as consuming meat or flying for leisure travel. To show that migration is special, we would need to show that the opportunity to migrate is sufficiently important that it should be socially guaranteed as a matter of what we owe each other.[55] Some have argued that migration is a human right, which would be one route to this conclusion.[56] But the claim that migration is a human right is controversial, and so this strategy would impose a significant justificatory burden on our response to the environmental argument. The claim that migration is a human right is also a premise that advocates of the environmental argument are unlikely to accept: If migration were a human right, then the environmental argument would not even get off the ground in the first place. This makes it unwise to hang our response to the environmental argument on the claim that migration is a human right.

The argument from human self-understanding is even less promising as a response to the environmental argument. Perhaps the opportunity to procreate should be viewed as a fundamental part of being treated as a human—I remain agnostic on that question here. But this is much less plausible in the case of migration. It is certainly true that humans are "a species that migrates and has developed many social practices and traditions that give meaning to migration." But this is much less plausible as an interpretation of what it means to live and be treated as a human. Migration is a central part of many cultures and traditions, but only a small minority of the global population (around 3.6% in 2022) is an international migrant.[57] Even in contexts in which migration is relatively unrestricted, such as the European Union, rates of international migration remain low. Unlike procreation, the vast majority of people in the world do not engage in international migration. This makes it hard to credibly claim that migration is a central part of human self-understanding. If this were true, it would imply that the vast majority of the global population does not live a characteristically human life.

The argument that migration is special does not provide a successful response to the environmental argument for immigration restrictions. Migration can certainly be a valuable project for those who engage in it, either because of its instrumental benefits or because of its non-instrumental value. But this alone does not suffice to show that it should be treated as special. Migration's instrumental value suggests that the opportunities that it allows the worst-off to access ought to be treated as special within climate mitigation, rather than migration itself. Even if migration is non-instrumentally valuable, this does not suffice to show that it should be protected. And the claim that migration is a central part of human self-understanding strains credibility in the face of the fact that most people do not migrate.

Migration and Development

The second counterargument to the environmental argument for immigration restrictions is made by Michael Ball-Blakely, who argues that it is inconsistent to promote immigration restrictions at home and human development abroad.[58] Ball-Blakely argues that advocates of the environmental argument face a dilemma: Either they must reject efforts to promote human development abroad, or they must reject the environmental argument for immigration restrictions. Since few would reject efforts to promote human development, he takes this to give us good reason to reject the environmental argument. My suggestion is that Ball-Blakey's argument is partially successful. Advocates of the environmental argument cannot reject human development efforts without violating egalitarian commitments. But they may be able to escape Ball-Blakely's dilemma, if they can explain why we should be optimistic about climate-friendly development but pessimistic about reducing emissions in high-income states.

Advocates of the environmental argument do typically accept that high-income states should promote human development efforts abroad. And if the environmental argument is to be made in a way that can appeal to egalitarians, then it is important for advocates of the environmental argument to uphold this commitment. Cafaro and Staples accept that high-income states have duties of global justice to the worst-off, but they suggest that these duties should be discharged through development aid, trade policies, and individual charitable donations, rather than through immigration policy.[59] In this way, they suggest that immigration restrictions can be consistent with the egalitarian goal of eliminating inequality and poverty.

Ball-Blakely argues that it is inconsistent for Cafaro and Staples to promote human development abroad and immigration restrictions at home.[60] His argument is that efforts to promote human development will *also* lead to an increase in greenhouse gas emissions, as emissions will tend to increase as the worst-off become increasingly affluent. This suggests that if high-income states should restrict immigration for environmental reasons, then they should also avoid investing in human development abroad for environmental reasons. Both migration and development lead to increased greenhouse gas emissions overall. As a result, "immigration and other poverty reduction strategies stand or fall together."[61] Ball-Blakely goes on to suggest that since we should not reject human development on environmental grounds, we should similarly not endorse immigration restrictions on environmental grounds.[62]

Ball-Blakely's argument depends on the empirical claim that efforts to promote human development will lead to increased greenhouse gas emissions, and on the moral claim that high-income states ought to resolve this dilemma by abandoning the environmental argument for immigration restrictions (rather than rejecting human development efforts abroad). Could advocates of the environmental argument push back against either of these claims?

One response to the dilemma that Ball-Blakely poses is simply to bite the bullet and reject the moral claim that high-income states should pursue human development efforts abroad. This is essentially the position taken by Garrett Hardin in what he calls the "lifeboat ethic."[63] But this position amounts to arguing that the worst-off should be sacrificed for the benefit of the better-off: It says that we should prevent the worst-off from accessing the benefits of human development, because their doing so would threaten the ability of the better-off to maintain their position of advantage. For this reason, the option of biting the bullet will strike many as morally reprehensible. It is also clearly incompatible with the egalitarian commitments that made the environmental argument an important challenge in the first place. So, if advocates of the environmental argument bite the bullet in the face of Ball-Blakely's dilemma, their argument is robbed of any force that it would have against the egalitarian who would otherwise be in favor of less restrictive immigration policies.

What about the empirical claim that promoting human development abroad will lead to increased greenhouse gas emissions? If low-income states follow the same development trajectories as high-income states, then we do have good reason to expect that development will lead to a significant increase in greenhouse gas emissions.[64] But this development pathway is not inevitable. The most recent Intergovernmental Panel on Climate Change (IPCC) report suggests that "eradicating extreme poverty, energy poverty, and providing decent living standards in low-emitting countries/regions in the context of achieving sustainable development objectives, in the near term, can be achieved without significant global emissions growth."[65] Achieving this goal is by no means straightforward and would require unprecedented forms of global cooperation and investments in climate-resilient development. But it is not inevitable that development will undermine climate mitigation.

However, even if climate-friendly development is possible, this does not save the environmental argument. After all, the reduction of greenhouse gas emissions in high-income states is *also* possible, if those states put in place aggressive climate mitigation policies. If advocates of the environmental argument are willing to be optimistic about climate-friendly development, then they should also be willing to be optimistic about emissions reductions in high-income states. Otherwise, they open themselves up to the objection that their optimism is selective. But optimism about reducing emissions in high-income states would undermine the environmental argument, since migration from low- to high-income states is only an environmental problem if emissions in high-income states remain high.

Advocates of the environmental argument could respond to Ball-Blakely's objection by explaining why we should be pessimistic about bringing down emissions in high-income states but optimistic about climate-friendly development in low-income states. If they could provide a reason for pessimism when it comes to emissions reductions in high-income states that does not apply to development in low-income states, then perhaps they could escape Ball-Blakely's dilemma. It is not clear what the prospects are for responding to Ball-Blakely's argument in this way, since at least at first glance it is unclear why we would be optimistic about climate-friendly development but not about emissions reductions in high-income states.[66] And in the absence of such an explanation, the dilemma does appear to undermine the environmental argument. So, even if it does not provide a conclusive reason to reject the environmental argument, Ball-Blakely's argument does put advocates of the environmental argument on the back foot.

Ball-Blakely's argument that it is inconsistent to advocate for immigration restrictions at home and development abroad is thus partially successful. Advocates of the environmental argument cannot bite the bullet and reject human development efforts without violating egalitarian commitments. They could attempt to reject the empirical basis of Ball-Blakely's argument by arguing that human development need not inevitably lead to increased greenhouse gas emissions. But the problem with this response is that migration from low- to high-income states will not inevitably lead to increased greenhouse gas emissions either, since high-income states could engage in aggressive climate mitigation. Advocates of the environmental argument cannot be pessimistic about migration while being optimistic about development, otherwise their pessimism appears selective.

Unless they can provide some explanation for why we should be pessimistic about reducing emissions in high-income states but not about climate-friendly development, then Ball-Blakely's provides a reason to reject the environment argument.

CLIMATE CHANGE AND THE JUSTIFICATION OF IMMIGRATION RESTRICTIONS

The final counterargument to the environmental argument for immigration restrictions says that high-emitting states cannot appeal to environmental arguments as a justification for exclusion. This argument starts from the idea that states that exclude owe would-be immigrants a justification for their exclusion. States cannot exclude for any reason whatsoever; they must provide some kind of justification for exclusion to those who they exclude.[67] And if exclusion is to be justified in terms that are acceptable to egalitarians, then the justification must take seriously the interests that would-be immigrants have in migrating to high-income states. The environmental argument aims to supply reasons that can justify the exclusion of would-be immigrants. My suggestion, however, is that states with high per capita emissions cannot appeal to those reasons to justify the exclusion of would-be immigrants.[68]

Anna Stilz has recently developed a model for assessing the justification of immigration restrictions that she calls the "conditional model" of immigration, which provides a useful way of evaluating the environmental argument.[69] The conditional model aims to balance the interests of would-be immigrants against the possible harms that might arise as a consequence of migration. On this model, states have a duty to permit migration unless such migration would involve significant harms. Stilz focuses on harms to the inhabitants of the receiving society and argues that states are entitled to give a degree of priority (though not unlimited priority) to the interests of their own members.[70] The conditional model is thus clearly not strongly egalitarian. But the environmental argument (at least in its global version) appeals to the harms of migration not for the inhabitants of the receiving society, but for those who are harmed by climate impacts more broadly. So, when applied to the environmental argument, the conditional model does not require us to weigh the interests of the receiving state's inhabitants more heavily than those of would-be immigrants, making it consistent with egalitarianism.

If we use the conditional model to assess the environmental argument, then one key question for our purposes is whether migration from low- to high-income states is harmful. If migration is not harmful, then states have a duty to permit it, since it serves important interests for would-be immigrants.[71] But if migration is harmful, then states may be permitted to restrict immigration. Not just *any* harm is sufficient to justify exclusion, at least if we understand a "harm" in the broad, non-moralized sense as a mere setback to an interest.[72] Otherwise, immigration restrictions could be justified whenever migration conflicted with idiosyncratic interests, such as the gourmand's interest in the existence of a sizeable market for their favorite cuisine, or perhaps even immoral preferences, such as the white nationalist's preference for excluding non-whites.[73] Instead, Stilz suggests that we should understand harms as "setbacks to certain legitimate moral interests that, as a matter of justice, ought to be protected."[74] This would rule out idiosyncratic interests and immoral preferences being used to justify immigration restrictions. As well as being harms in this more robust and moralized sense, harms that justify immigration restrictions cannot be trivial: Harms associated with migration need to be sufficiently serious, if they are to outweigh the interests of would-be immigrants and justify immigration restrictions.

The environmental argument does seem to give us good reason to think that migration from low- to high-income states is harmful in this sense. After all, if the empirical premises in the argument are correct, then they suggest that migration will tend to increase greenhouse gas emissions, exacerbating the negative consequences of climate change for the worst-off. The impacts of climate change are clearly significantly harmful to fundamental human interests: They lead to illness and premature death, undermine livelihoods and exacerbate poverty, threaten access to food and water, destroy essential infrastructure and lead to displacement, and more besides.[75] These impacts disproportionately affect those who are already among the worst-off. And the more we emit, the worse those impacts will become over time. So, climate impacts are clearly the right kind of harm to justify immigration restrictions. And as we have already seen, migration from low- to high-income states does appear to increase greenhouse gas emissions overall—or, at least, this is a plausible empirical claim that I do not contest here. Advocates of the environmental argument thus appear to have a justification for exclusion that satisfies the conditional model. High-income states can justify exclusion to would-be immigrants by arguing that the predictable consequence of admission would be an increase in greenhouse gas emissions that would, ultimately, lead to significant harms to the worst-off.

The problem with this justification is not that migration from low- to high-income states does not have harmful consequences. It is rather that the consequences of migration are only harmful because of the actions of the receiving state. Migration from low- to high-income states may well increase overall contributions to climate change. But it does so only because per capita emissions within high-income states remain at a higher level than can be sustained globally. As we saw earlier, migration does not increase population levels overall, only the relative share of the global population within high-income states. This means that if high-income states were to bring down their per capita emissions to a globally sustainable rate, then migration would not have harmful environmental consequences.

In the first instance, this suggests that high-income states ought to adopt climate mitigation policies that make migration less environmentally harmful. If high-income states were to reduce their per capita emissions, then there would be no conflict between the interests of would-be immigrants and the interests of those who are vulnerable to climate change. High-income states have it in their power to *make* migration consistent with tackling climate change, by adopting aggressive climate mitigation policies that bring down their per capita emissions. If it were to do so, then migration would no longer be harmful, and the state would no longer have any environmental reason to exclude would-be immigrants. In a discussion of economic migration, Stilz suggests that migration may have negative economic consequences for some incumbent residents, but that those costs can be avoided if liberalized immigration policies are paired with the right social and economic policies.[76] Similarly, in the context of climate change, liberalized immigration policies can be paired with aggressive climate mitigation in order to avoid the environmental harms associated with migration. Egalitarians should thus push high-income states to adopt aggressive climate mitigation policies, not only because such policies limit the negative consequences of climate change for the worst off, but also because such policies make the egalitarian goal of loosening immigration restrictions in high-income states consistent with climate mitigation.

Migration only has harmful environmental consequences because high-income states are *failing* to do what justice requires of them: to radically reduce their per capita emissions.[77] The environmental argument appeals to what Stilz calls a "non-ideal" reason for immigration restrictions: a "reason that we have in virtue of the fact that other agents are poorly motivated, and their unwillingness to comply with their moral duties threatens significantly harmful social consequences."[78] But in the case of the environmental argument, the agent that is acting unjustly is the very same agent that is seeking to restrict immigration. The state is not seeking to restrict immigration to protect the victims of climate change against the unjust actions of *others*, it is seeking to restrict immigration to protect the victims of climate change against its *own* unjust actions in refusing to reduce its emissions to sustainable levels.

This means that high-emitting states cannot themselves appeal to the environmental argument to justify immigration restrictions. Johann Frick has pointed out that justification is "speaker-relative," which is to say the identity of the agent who is justifying an action can make a difference to whether or not their justification is successful.[79] In a famous case, G. A. Cohen points out that a kidnapper cannot appeal to their own willingness to kill their victim as a moral justification for why the ransom ought to be paid.[80] It might be true, as a matter of fact, that the kidnapper will kill their victim if the ransom is not paid. But this is not something to which the kidnapper can appeal as a justification. After all, the kidnapper is perfectly able to release the victim without being given the ransom; it is only the kidnapper's refusal to do what morality requires of them that makes it the case that the victim cannot be released without the ransom being paid. The kidnapper cannot appeal to their own refusal to do what morality requires of them as a moral justification.

In the same way, high-emitting states cannot appeal to the environmental argument as a justification for the exclusion of would-be immigrants. After all, it is only because states are failing to do what justice requires of them—reduce their per capita emissions to a sustainable rate—that migration has harmful environmental consequences in the first place. High-income states could perfectly well enact aggressive climate mitigation policies that brought down their per capita emissions to a sustainable level. It may be true that migration from low- to high-income states will, in fact, lead to increased greenhouse gas emissions if high-income states do not engage in aggressive climate mitigation. But high-income states cannot appeal to this as a justification for exclusion: They cannot appeal to their own unjust behavior as a reason to exclude would-be immigrants.

A sincere commitment to tackling climate change would require high-income states to significantly reduce their per capita emissions. If they fail to do so, they cannot at the same time appeal to the environmental argument as a justification for the exclusion of would-be immigrants. But if high-income states *were* to radically reduce their per capita emissions—as justice requires of them— then they would not need to restrict immigration in the first place. High-income states thus cannot appeal to the environmental argument in either scenario. Either high-income states radically reduce their emissions, in which case they do not need to restrict immigration, or they do not reduce their emissions, in which case they cannot appeal to the environmental argument as a justification for immigration restrictions. So, if we accept the minimal premise that states that restrict immigration owe a justification for exclusion to would-be immigrants, then the environmental argument cannot justify immigration restrictions.

Conclusion

The environmental argument for immigration restrictions suggests that egalitarians who are concerned about climate change ought to support restrictive immigration policies in high-income states. If my arguments are correct, then egalitarians need not endorse this conclusion. Advocates of the environmental argument may well be correct to suggest that at present migration from low- to high-income states leads to increased greenhouse gas emissions overall, but this does not mean that high-income states should restrict immigration. This is not because migration is in some way special, such that migration-related emissions should be exempted from climate mitigation efforts. Nor is it primarily because it is inconsistent to promote development abroad while promoting immigration restrictions at home, though it may well be inconsistent. Rather, it is because high-income states are in no position to avail themselves of the environmental argument for restricting immigration. If they were genuinely committed to tackling climate change, then they would enact aggressive climate mitigation policies that brought down their emissions to a globally sustainable rate. And if they were to do so, then there would no longer be any environmental reason to restrict immigration in the first place.

NOTES

1 Intergovernmental Panel on Climate Change (IPCC), "Summary for Policymakers," in *Climate Change 2023: Synthesis Report. Contribution of Working Groups I, II and III to the Sixth Assessment Report of the Intergovernmental Panel on Climate Change*, ed. Core Writing Team, Hoesung Lee, and José Romero (2023), 5.

2 Branko Milanovic, *Global Inequality: A New Approach for the Age of Globalization* (Cambridge, MA: Harvard University Press, 2016), 118–154; see also Gordon H. Hanson, "The Economic Consequences of the Migration of Labor," *American Economic Review* 1, no. 1 (2009): 179–208; Michael A. Clemens, Claudio E. Montenegro, and Lant Pritchett, "The Place Premium: Bounding the Price Equivalent of Migration Barriers," *Review of Economics and Statistics* 101, no. 2 (2019): 201–213.

3 For climate mitigation, see Darrel Moellendorf, *Mobilizing Hope: Climate Change and Global Poverty* (Oxford: Oxford University Press, 2022); Simon Caney, "Climate Change and the Duties of the Advantaged," *Critical Review of International Social and Political Philosophy* 13, no. 1 (2010): 203–238; Henry Shue, *Climate Justice: Vulnerability and Protection* (Oxford: Oxford University Press, 2014). For immigration, see Joseph Carens, *The Ethics of Immigration* (Oxford: Oxford University Press, 2013), 233–236; Darrel Moellendorf, *Cosmopolitan Justice* (New York: Routledge, 2002), 61–67; Nils Holtug, "Global Equality and Open Borders," in *Oxford Studies in Political Philosophy Volume 6*, ed. David Sobel, Peter Vallentyne, and Steven Wall (Oxford: Oxford University Press, 2020).

4 Different versions of this argument include Philip Cafaro and Winthrop Staples III, "The Environmental Argument for Reducing Immigration into the United States," *Environmental Ethics* 31, no. 1 (2009): 5–30;

Philip Cafaro, *How Many Is Too Many? The Progressive Argument for Reducing Immigration into the United States* (Chicago: University of Chicago Press, 2015), 105–176; Robert H. Chapman, "Immigration and Environment: Setting the Moral Boundaries," *Environmental Values* 9, no. 2 (2000): 189–209; and David Miller, *Strangers in Our Midst: The Political Philosophy of Immigration* (Cambridge, MA: Harvard University Press, 2016), 65–66.

5 Jamie Draper, *Climate Displacement* (Oxford: Oxford University Press, 2023); Jamie Draper, "Climate Change and Displacement: Towards a Pluralist Approach," *European Journal of Political Theory* 23, no. 1 (2024): 44–64.

6 Joe Turner and Dan Bailey, "'Ecobordering': Casting Immigration Control as Environmental Protection," *Environmental Politics* 31, no. 1 (2021): 110–131.

7 See John Hultgren, *Border Walls Gone Green: Nature and Anti-Immigrant Politics in America* (Minneapolis: University of Minnesota Press, 2015).

8 Cafaro and Staples, "The Environmental Argument."

9 Cafaro and Staples, "The Environmental Argument," 5–6. If we want to allow for the possibility that countervailing reasons could outweigh environmental reasons, then strictly speaking the conclusion of the argument should refer to a *pro tanto* moral duty, but Cafaro and Staples do not restrict their argument in this way.

10 Cafaro and Staples, "The Environmental Argument," 6.

11 Cafaro and Staples, "The Environmental Argument," 9–11.

12 Cafaro and Staples, "The Environmental Argument," 11–12.

13 Miller, *Strangers in Our Midst*, 65–66.

14 Miller, *Strangers in Our Midst*, 66.

15 For this distinction, see David Schlosberg, *Defining Environmental Justice: Theories, Practices, Movements* (Oxford: Oxford University Press, 2007), 3–8.

16 Cafaro and Staples, "The Environmental Argument," 15.

17 Cafaro and Staples, "The Environmental Argument," 16 n19.

18 Cafaro, *How Many Is Too Many?*, 161.

19 Cafaro and Staples, "The Environmental Argument," 25.

20 Cafaro and Staples, "The Environmental Argument," 21, 28. In support of the empirical part of this claim, Cafaro and Staples refer to George Borjas, *Heaven's Door: Immigration Policy and the American Economy* (Princeton, NJ: Princeton University Press, 1999). But Borjas's view by no means represents the consensus view among economists. For an alternative, see David Card, "Immigration and Inequality," *American Economic Review* 99, no. 2 (2009): 1–21. For a good parsing of the empirical literature and its relevance for philosophical arguments about immigration restric-

tions, see Anna Stilz, "Economic Migration: On What Terms?," *Perspectives on Politics* 20, no. 3 (2022): 983–998.

21 Cafaro and Staples, "The Environmental Argument," 15–16. Cafaro and Staples refer to a number of authors whose views suggest a broad interpretation of this non-anthropocentric ethic, including those of Holmes Rolston III, J. Baird Callicott, Bryan Norton, Arne Næss, Karen Warren, Val Plumwood, and Ronald Sandler.

22 See, for example, Moellendorf, *Mobilizing Hope*, 199–201; Darrel Moellendorf, *The Moral Challenge of Dangerous Climate Change: Values, Poverty, and Policy* (Cambridge: Cambridge University Press, 2014), 59–61.

23 This makes the claim involved in the ecological argument closer to something like Holmes Rolston III's controversial claim that saving nature should sometimes take priority over feeding the poor. See Holmes Rolston III, "Feeding People versus Saving Nature?" in *World Hunger and Morality*, ed. William Aiken and Hugh LaFollette (Englewood Cliffs, NJ: Prentice-Hall, 1996), 248–267. For critiques of Rolston's view, see Robin Attfield, "Saving Nature, Feeding People, and Ethics," *Environmental Values* 7, no. 3 (1998): 291–304; Alan Carter, "Saving Nature and Feeding People," *Environmental Ethics* 26, no. 4 (2004): 339–360.

24 The *locus classicus* is Paul R. Ehrlich and John P. Holdren, "A Bulletin Dialogue on 'The Closing Circle': Critique," *Bulletin of the Atomic Scientists* 28, no. 5 (1972): 16–27.

25 For the significance of this in the context of climate change, see Kenneth Gillingham, Richard G. Newell, and William A. Pizer, "Modelling Endogenous Technological Change for Climate Policy Analysis," *Energy Economics* 30, no. 6 (2008): 2734–2753.

26 The former idea is referred to as the "Jevons Paradox" or the "rebound effect"; see Richard York and Julius Alexander McGee, "Understanding the Jevons Paradox," *Environmental Sociology* 2, no. 1 (2016): 77–87. The latter idea is referred to as the "environmental Kuznets curve"; see Gene M. Grossman and Alan B. Kreuger, "Economic Growth and the Environment," *Quarterly Journal of Economics* 110, no. 2 (1995): 353–377. Evidence for the environmental Kuznets curve appears to be mixed at best when it comes to CO_2 emissions; see Marzio Galeotti, Alessandro Lanza, and Francesco Pauli, "Reassessing the Environmental Kuznets Curve for CO_2 Emissions: A Robustness Exercise," *Ecological Economics* 57, no. 1 (2006): 152–163.

27 Jochen Mayer and Regina T. Riphahn, "Fertility Assimilation of Immigrants: Evidence from Data Count Models," *Journal of Population Economics* 13, no. 2 (2000): 241–261; Holger Stitchnoth and Mustafa Yeter, "Cultural Influences on the Fertility Behavior of First- and Second-Generation Immigrants," *Journal of Demographic Economics* 82, no. 3 (2016): 281–314.

28 Stein Emil Vollset et al., "Fertility, Mortality, Migration, and Population Scenarios for 195 Countries and Territories from 2017 to 2100: A Forecasting Analysis for the Global Burden of Disease Study," *The Lancet* 396, no. 10258 (2020): 1285–1306. There may be offsetting factors that affect overall population levels, such as differences in life expectancies in low- and high-income states.

29 Sia Liang et al., "CO_2 Emissions Embodied in International Migration from 1995–2015," *Environmental Science & Technology* 54, no. 19 (2020): 12530–12538.

30 Liang et al., "CO_2 Emissions Embodied in International Migration," 12535.

31 Guizhen Ma and Erin Trouth Hoffman, "Population, Immigration, and Air Quality in the USA: A Spatial Panel Study," *Population and Environment* 40, no. 3 (2019): 283–302; Jay Squalli, "Disentangling the Relationship Between Immigration and Environmental Emissions," *Population and Environment* 43, no. 1 (2021): 1–21.

32 Draper, *Climate Displacement*, 86–105; see also Jamie Draper, "Labor Migration and Climate Change Adaptation," *American Political Science Review* 116, no. 3 (2022): 1012–1024.

33 Hélène Benveniste, Michael Oppenheimer, and Marc Fleurbaey, "Effect of Border Policy on Exposure and Vulnerability to Climate Change," *Proceedings of the National Academy of Sciences* 117, no. 43 (2020): 26692–26702.

34 Moellendorf, *Mobilizing Hope*, 140–151; Daniel Callies and Darrel Moellendorf, "Assessing Climate Policies: Catastrophe Avoidance and the Right to Sustainable Development," *Politics, Philosophy & Economics* 20, no. 2 (2021): 127–150. For an even more optimistic view, see Joseph Heath, *Philosophical Foundations of Climate Change Policy* (Oxford: Oxford University Press, 2021), 63–108.

35 Shue makes this argument with respect to Bioenergy with Carbon Capture and Storage (BECCS) technologies. See Henry Shue, "Climate Dreaming: Negative Emissions, Risk Transfer, and Irreversibility," *Journal of Human Rights and the Environment* 8, no. 2 (2017): 203–216.

36 For example, Tim Jackson, *Prosperity Without Growth: Foundations for the Economy of Tomorrow* (New York: Routledge, 2017), esp. 84–102; Jason Hickel and Giorgos Kallis, "Is Green Growth Possible?," *New Political Economy* 25, no. 4 (2020): 469–486. For a related discussion, see also Lucas Stanczyk, "On the Moral Challenge of the Climate Crisis," in this volume.

37 For an analysis of these disagreements, especially as they arise in ecological and environmental economics, see Mark Sagoff, "Carrying Capacity and Ecological Economics," *BioScience* 45, no. 9 (1995): 610–620.

38 Cafaro and Staples, "The Environmental Argument," 12, 23.
39 These data are drawn from Hannah Ritchie et al., "Population Growth" (2023), available at https://ourworldindata.org (accessed May 4, 2024) and Max Roser, "Fertility Rate" (2024), available at https://ourworldindata.org (accessed May 4, 2024).
40 Dudley Kirk, "Demographic Transition Theory," *Population Studies* 50, no. 3 (1996): 316–387.
41 Cafaro and Staples, "The Environmental Argument," 14–15.
42 Cafaro and Staples, "The Environmental Argument," 23.
43 See, for example, Elizabeth Cripps, "Climate Change, Population, and Justice: Hard Choices to Avoid Tragic Choices," *Global Justice: Theory, Practice, Rhetoric* 8, no. 1 (2015): 1–22; Clare Heyward, "A Growing Problem? Dealing with Population Increases in Climate Justice," *Ethical Perspectives* 19, no. 4 (2012): 703–732.
44 Ingrid Robeyns, "Is Procreation Special?," *Journal of Value Inquiry* 56, no. 4 (2022): 643–661.
45 Robeyns, "Is Procreation Special?," 643–644. For a defense of the moral equivalence thesis, see Thomas Young, "Overconsumption and Procreation: Are They Morally Equivalent?," *Journal of Applied Philosophy* 18, no. 2 (2001): 183–192.
46 Robeyns, "Is Procreation Special?," 648–656.
47 Robeyns, "Is Procreation Special?," 654–656.
48 Robeyns, "Is Procreation Special?," 656–659.
49 Robeyns, "Is Procreation Special?," 657.
50 Robeyns, "Is Procreation Special?," 658.
51 For a capabilities view of migration, see Hein de Haas, "A Theory of Migration: The Aspirations-Capabilities Framework," *Comparative Migration Studies* 9, no. 8 (2021): 1–35. Strictly speaking, migration itself is a *functioning*, while the effective ability or opportunity to engage in migration is its corresponding *capability*. For this distinction, see Ingrid Robeyns, *Wellbeing, Freedom, and Social Justice: The Capability Approach Re-Examined* (Cambridge: Open Book Publishers, 2017), 38–41.
52 Jonathan Wolff and Avner de Shalit, *Disadvantage* (Oxford: Oxford University Press, 2007), 133–136.
53 Valeria Ottonelli and Tiziana Torresi, "Inclusivist Egalitarian Liberalism and Temporary Migration: A Dilemma," *Journal of Political Philosophy* 20, no. 2 (2012): 202–224.
54 For a discussion of immigration and distributive justice that takes this idea as its starting point, see Eric Cavallero, "An Immigration-Pressure Model of Global Distributive Justice," *Politics, Philosophy & Economics* 5, no. 1 (2006): 97–127.
55 Robeyns, "Is Procreation Special?," 652.

56 See Kieran Oberman, "Immigration as a Human Right," in *Migration in Political Theory: The Ethics of Movement and Membership*, ed. Sarah Fine and Lea Ypi (Oxford: Oxford University Press, 2016), 32–56. For the contrary view, see David Miller, "Is There a Human Right to Immigrate?" in *Migration in Political Theory: The Ethics of Movement and Membership*, ed. Sarah Fine and Lea Ypi (Oxford: Oxford University Press, 2016), 11–31.

57 Marie McAuliffe and Anna Triandafyllidou, eds., *World Migration Report 2022* (International Organization for Migration [IOM], 2021), 21; see also Hein de Haas, *How Migration Really Works* (New York: Viking, 2023), 16–19.

58 Michael Ball-Blakely, "Climate Change and Green Borders: Why Closure Won't Save the World," *Philosophy in the Contemporary World* 28, no. 2 (2022): 70–95. For related discussions, see Peter Higgins, *Immigration Justice* (Edinburgh: Edinburgh University Press, 2013), 93–96; Chandran Kukathas, *Immigration and Freedom* (Princeton, NJ: Princeton University Press, 2021), 153–155.

59 Cafaro and Staples, "The Environmental Argument," 22–23.

60 Ball-Blakely, "Climate Change and Green Borders."

61 Ball-Blakely, "Climate Change and Green Borders, 81."

62 Ball-Blakely, "Climate Change and Green Borders, 82."

63 Garrett Hardin, "Living in a Lifeboat," *BioScience* 24, no. 10 (1972): 561–568.

64 For a discussion of the implications of this for global climate policy, see Moellendorf, *The Moral Challenge of Dangerous Climate Change*, 132–134.

65 IPCC, "Summary for Policymakers," 30.

66 Ball-Blakely suggests that, if anything, we have better grounds for optimism about emissions reductions in high-income states than for climate-friendly development, because high-income states have the means to access clean energy technology. Ball-Blakely, "Climate Change and Green Borders," 82.

67 See Joseph Carens, *The Ethics of Immigration* (Oxford: Oxford University Press, 2013), 174–175; Miller, *Strangers in Our Midst*, 104–5; Anna Stilz, *Territorial Sovereignty: A Philosophical Exploration* (Oxford: Oxford University Press, 2019), 212–213.

68 For related arguments, see Kukathas, *Immigration and Freedom*, 153–155, and Ryan Pevnick, *Immigration and the Constraints of Justice* (Cambridge: Cambridge University Press, 2011), 151–152. Chris Bertram also makes a version of this argument, although he focuses on the exclusion of people whose migration is at least partly driven by climate change (manuscript on file with the author).

69 Stilz, *Territorial Sovereignty*, 187–215.

70 Stilz, *Territorial Sovereignty*, 187–189.

71 Stilz, *Territorial Sovereignty*, 188.

72 See Joel Feinberg, *The Moral Limits of Criminal Law, Vol. 1: Harm to Others* (Oxford: Oxford University Press, 1981), 33–34.

73 I set aside the question about whether such preferences could count as genuine interests here.

74 Stilz, *Territiorial Sovereignty*, 187–188.

75 IPCC, "Summary for Policymakers," 6–7.

76 Anna Stilz, "Economic Migration: On What Terms?"

77 For defenses of the view that the state is the relevant moral agent for climate mitigation, see Darrel Moellendorf, "Taking UNFCCC Norms Seriously," in *Climate Justice in a Non-Ideal World*, ed. Clare Heyward and Dominic Roser (Oxford: Oxford University Press, 2016) and Blake Francis, "In Defence of National Climate Responsibility: A Reply to the Fairness Objection," *Philosophy & Public Affairs* 49, no. 2 (2021): 115–155.

78 Stilz, *Territorial Sovereignty*, 196.

79 Johann Frick, "What We Owe to Hypocrites: Contractualism and the Speaker-Relativity of Justification," *Philosophy & Public Affairs* 44, no. 4 (2016): 223–265.

80 G. A. Cohen, *Rescuing Justice and Equality* (Cambridge, MA: Belknap Press, 2008), 38–41.

INDEX

Page numbers in italics indicate Figures

Abramovich, Roman, 38
academia, uniformity in, 73
Acid Rain Trading program, US, 151
adaptation: can-kicking of, 224; global temperature relationship with, 23–25; immigration relationship with, 273; mitigation trade off with, 188, 189
Africa: fossil fuel industry projects in, 187; Sub-Saharan, 260n41
agency: in can-kicking, 230–33; of consumer choices, 103–4; fragmented, 99. *See also* moral agency
aggregation problem, 82
air conditioning problem, 188
air pollution: capitalism and, 129; regulation of, 129, 151
Akbar, Amna, 252–53
akrasia (failure to act on one's judgment), 123n123
albedo cooling effect, 171
American Petroleum Institute, 151
American Society for Political and Legal Philosophy, 1
AMOC. *See* Atlantic meridional overturning circulation
Anarchy, State, and Utopia (Nozick) (1974), 86
animals: animal-regarding approach, 186; social welfare costs of, 196n36
antitrust law, 142, 144, 146
Arctic, forest fires in, 165
Arrow, Kenneth, 195n29
asset managers, 148
Atlantic meridional overturning circulation (AMOC), 191n5
atomic age, 152

authoritarianism, of China, 130–31
auto-catalytic social effects, of climate change, 188, 251
aviation emissions, 39, 174, 192n9, 205–7

Baker, Shalanda, 253
Ball-Blakely, Michael: on emissions of high-income states, 292n66; on immigration restrictions and human development, 278–80, 281–82
Battistoni, Alyssa, 3, 148–49, 157n26; capitalism critiqued by, 4, 132–37, 151; on consumer responsibility, 142, 146; on externalities, 45, 141; on harm, 158n41; on "The Problem of Social Cost," 145–46; on social costs, 157n25; standard economic environmental policy critiqued by, 132–37; Young informing, 161n75
BECCS. *See* Bioenergy with Carbon Capture and Storage technologies
Benveniste, Hélène, 273
Bertram, Chris, 292n68
Biden administration: can-kicking under, 237; on social cost of carbon, 26
Bidenomics, 150
Bioenergy with Carbon Capture and Storage technologies (BECCS), 290n35
Black Lives Matter protests, 160n60
board of directors: diversity of, 160n67; shareholder primacy for, 147–48
Borjas, George, 288n20

295

Bratton, William, 145
Broome, John, 232
Budolfson, Mark, 4, 118n54, 124n132, 124n135, 194n27
Bullard, Robert, 153
burden dumping, 227
burden-sharing, mitigation, 225, 229, 231, 240
business-as-usual development, 30–31

Cafaro, Philip, 288n20; on consumption growth, 173, 174–75, 279; immigration restriction arguments by, 266, 267, 268, 269, 279, 288n9; on intergenerational justice, 268; nonanthropocentric ethic of, 289n21; on population growth, 274; on responsibility, 279
Campaign to Make General Motors Responsible, 152–53
Canada: fossil fuel industry projects in, 169, 187–88; politicians in, 169
can-kicking, 7, 227–28, 240; of adaptation, 224; agency in, 230–33; carbon budget shares and, 229–30, 236–37; ethics of, 238–39; intergenerational injustice of, 6, 167–68, 184, 204; of mitigation, 223, 224, 233, 236, 238; responsibility for, 225, 230–33; slack-taking compared to, 233; Stanczyk on, 226; of states and individuals, 231–32; varieties of, 236–39
cap-and-trade programs, 109
capitalism: Battistoni critiquing, 4, 132–37, 151; Cohen on, 94; democracy compared to, 130, 131; domination under, 4, 133, 134–35, 151; externalities and, 111–12, 129, 137; Marx on, 80; politics of, 167; poor persons within, 134; regulation within, 4, 128–29, 130, 131, 134, 135, 137; structural domination of, 94–96; unfreedom within, 98, 108–9. *See also* well-regulated capitalism
carbon: counterfactual, 27–31; super-spreaders, 35–37, *37*, 38, *38*, 39

carbon budgets, 6, 166; can-kicking and, 229–30, 236–37; individual, 235–36, 239; for intergenerational justice, 7; for 1.5 degrees Celsius warming, 48, 171; stock-taking of, 235–36; for 2 degrees Celsius warming, 171–72; US, 234, 235
carbon capture, 159n50, 191n6; corporations supporting, 258n25; development of, 189; IPCC models on, 245
carbon emissions, 92, 101, 189; of corporations, 109; in Democratic Republic of Congo, 247; of elite persons, 82, 202; GDP and population growth relationship with, 172, 192n11; income correlation with, 98; individual, 98; inequality relationship with, 40, 72, 244; reforestation relationship with, 174; US, 234, 247; welfare economic approach to, 51
carbon footprints: collective, 230; individual, 230, 235–36
carbon inequalities, within and between countries, 36–37
carbon-intensive behaviors, 42
carbon markets, 4; carbon offset relationship with, 28, 30–31; carbon upsets compared to, 32
carbon offsets: carbon market relationship with, 28, 30–31; carbon upsets compared to, 33; for children, 34; counterfactual representations of, 29–30
carbon removal: IPCC model on, 245, 257n22; Switzerland on, 246
carbon taxes: within capitalism, 129; fossil fuel industry on, 110–11; IPCC proposals for, 91, 109–10; in US, 126
carbon upsets, 3, 8, 31; carbon markets compared to, 32; carbon offsets compared to, 33; digital art exhibition on, 58n61; politics of, 32–33, 51, 58n61
Chicago School, 154; on antitrust law and prices, 146; on common good, 149; on "The Nature of the Firm,"

Index

142, 143, 144, 149; on transaction costs, 143, 151–52
children, carbon offsets for, 34
China: authoritarianism of, 130–31; coal in, 139n17, 169; household emission level disparities within, 39
class society: *Climate Change as Class War* on, 250–51; inequality of, 93; structural domination and social costs within, 90–98
Clean Air Act (US), 15, 26, 50
Clean Development Mechanism, 28
Clean Power Plan (US), 49, 50, 238
climate change. *See specific topics*
Climate Change as Class War (Huber), 250–51
climate economics, 15, 44, 53n10
climate ethics, 169; in politics, 7, 8, 243; stock-taking in, 225
climate investments, 257n21
climate justice. *See* environmental justice
climate policy, 7, 17, 44, 136; cost-benefit analysis of, 14, 18; expert knowledge and, 71–74; GHG responsibility stalling, 27–28; on immigration, 248; industrial, 251, 253; inequality reduction as, 67–71; Kapp on, 110; neoliberal hegemony in, 16; Stanczyk on, 243
climate policy, US: climate economics influencing, 53n10; extraterritorial effects considered by, 27, 55n27; regulation, 15; on social cost of carbon, 26–27. *See also* Inflation Reduction Act
climate scientists, 3
climate tipping points, 170–71, 186, 250
climate treaty, Posner and Weisbach on, 170
climate vulnerability: from immigration restrictions, 273; structural domination of, 95–96
Climework, 197n51

coal: in China, 139n17, 169; energy demand for, 187; energy vulnerable households scavenging, 244, 250; LNG compared to, 196n41; sulfur pollution from, 171
Coase, Ronald, 109, 117n51; corporate law based on, 142, 144, 147; on domination, 141; Kapp on, 91; on nature and politics, 89; Pigou compared to, 86–90; on pollution rights, 91–92, 96, 112; "The Problem of Social Cost" by, 86–88, 141, 145–46; on social costs, 86–88, 90, 110, 141; on state intervention, 148–49. *See also* "The Nature of the Firm"
Coase Theorem, 88
cobalt, 192n9
Cohen, G. A.: kidnapping analogy of, 285; on labor and capitalism, 94
colonized peoples, formerly, 4
commodity fetishism, 100
common good, 148–49
conditional model, of immigration, 282–84
Condon, Madison, 4–5, 83; on externalities, 119n66; on state intervention, 110
consequentialism: rights, 205–15; Stanczyk rejecting, 205; utilitarian form of, 199–200. *See also* nonconsequentialism
Constitution, US, 50–51
consumer choices, 4; agency of, 103–4; complicity of, 99–101
consumer responsibility, 99–101; Battistoni on, 142, 146; for intergenerational justice, 203–4
consumer sovereignty, 161n73
consumption-based emissions, 7
consumption growth: Cafaro on, 173, 174–75, 279; IPAT formula on, 270–71; of meat, 173–74; population growth and, 173–78, 193n21, 220n7, 242, 247–48; Stanczyk on, 220n7, 242, 243–44, 252; Staples on, 274

contestation: of modeling assumptions, 25, 68; of scientific models, 3
co-pollutants, 30
corporate consolidation, 144
corporate law: Coasean framing in, 142, 144, 147; decarbonization through, 147
corporate responsibility, 5, 109, 142
corporations: carbon capture support of, 258n25; carbon emissions of, 109; IRA tax credits for, 263n74; Jensen and Meckling on, 145, 147; Means on, 156n15; "The Nature of the Firm" on, 143; nexus of contracts theory of, 144–45, 146, 156n19; shareholder primacy of, 5, 142; Young on, 153. *See also* externalities
cost-benefit analysis: assumptions underlying, 20, 25; of climate policy, 14, 18; critiques of, 66, 67–68; of decarbonization, 147; equity weighting within, 139n21; regulatory, 13. *See also* discounted future cost-benefit analysis
cost-benefit analysis, US, 56n37; under Obama, 18; of Obama task force, 22; regulatory, 18, 21–22, 25–26, 27; on social cost of carbon, 19–20, 27; under Trump administration, 25–26, 27
Cripps, Elizabeth, 232
Critique of Dialectical Reason (Sartre) (1960), 106

DAC. *See* Direct Air Capture technology
Daly, Herman, 47, 48–49
decarbonization, 44, 83, 159n47, 167; of air conditioning energy infrastructure, 188; through corporate law, 147; in Europe, 169; of Southern Company, 246; state intervention for, 110
decent-living emissions: of energy vulnerable households, 244–45, 250, 256n17; luxury emissions compared to, 244, 245

"Default Libertarian" approach, 118n54
degrowth, 48, 67; inequality and emissions reduction relationship with, 69; knowledge production of, 74; moral arguments of, 72–73; on technological development, 274
dehumanization, of procreation denial, 276
Delayed-Action Bomb case, intergenerational justice of, 216–17, 220n11
democracy: authoritarianism, 130–31; capitalism compared to, 130, 131; excess, 131, 139n19; scientific experts and, 3, 67, 71–74
Democratic Republic of Congo, per capita carbon emissions in, 247
deontology, Kamm on, 6
depoliticization, 3, 153
diagnostic question, 2
DICE model, 19, 46; on GDP and global temperature, 23–24, 24, 25, 49, 63n123, 175; Nordhaus on, 49
Direct Air Capture technology (DAC), 197nn51–52
discounted future cost-benefit analysis, 66, 177, 257n21; Obama task force on, 19, 22; regulatory, 22–23; Tribe on, 13; of Trump administration, 25; US using, 18
disjunctive wrongs, 226, 227; on harm, 183, 224; Stanczyk applying, 224
distributive justice, 229; within capitalism, 129; of environmental degradation reduction, 267; migration and, 277; of neoclassical economics, 44–45; uniform disincentives ignoring, 127
domination, 82; under capitalism, 4, 133, 134–35, 151; Coase on, 141; in politics, 97–98, 112, 245; responsibility of elite persons for, 133, 134, 253; social, 83; of weather, 124n132, 135. *See also* market domination; structural domination

Index

double movement (Polanyi), 91
Draper, Jamie, 8

Earth Day (1970), 152, 160n60
Eclipse (yacht), 38
economic inequality: emissions gap resulting from, 36–37, *37*, 38–39, 40, 41, 59n76, 82; environmental justice, 35–36; GHG relationship with, 41; Law and Political Economy movement on, 252; mitigation addressing, 42, 66. *See also* elite persons
economics: empty-world and full-world, 47; politics relationship with, 80
The Economics of Welfare (Pigou) (1920), 84
economists, policymaking of, 126
economy: carbon-dependent, 256n19; environment relationship with, 46–48; GHGs decoupled from, 48; global, 172–73; immigration relationship with, 284; population growth relationship with, 260n39; scale of, 47
egalitarians: on conditional model, 282–83; on human development, 279, 280; on immigration, 264; on immigration restrictions, 264, 265–66, 269, 270, 282, 285, 286; on mitigation, 264, 285
EJ. *See* environmental justice
electric vehicles, 192n9
elite persons: carbon emissions of, 82, 202; carbon superspreaders, 35–37, *37*, 38, *38*, 39; environmental justice worsened by, 190; household emission levels, 37, 39, 42, 59n72; Latour on, 154; luxury consumption of, 202; luxury emissions of, 245, 258n31; moral failures of, 261n62; politics influenced by, 245; responsibility of, 133, 134, 154–55, 253; Stanczyk on, 202, 245, 251, 255
emissions: aviation, 39, 174, 192n9, 205–7; in China, 39; GDP relationship with, 192n11; of high-income states, 272, 275, 281–82, 284, 286, 292n66; from immigration, 275; intergenerational injustice of, 180–81; migration-related, 265; models relating, 42; population growth relationship with, 174, 192n11, 193n19, 260n41; subsistence, 35, 244. *See also* greenhouse gas emissions
emissions gap: from economic inequality, 36–37, *37*, 38–39, 40, 41, 59n76, 82; from inequality, 59n76
emissions reduction: degrowth relationship with, 69; equality and, 52, 69; from gender equality, 43, 72; utilitarian approach to, 176
energy demand: for coal, 187; of DAC, 197n51; per capita, 167
energy infrastructure, 188, 236
energy investment decisions, 21
energy vulnerable households, decent-living emissions of, 244–45, 250, 256n17
environmental degradation: distributive justice for, 267; immigration restriction argument based on, 264–70; moral responsibility for, 266–68
environmental impact, population, affluence, and technological development (IPAT), 270–71
environmental justice (EJ), 30, 81–82; economic inequality, 35–36; elite persons worsening, 190; against standard economic environmental policy, 127; against uniform disincentives, 128; on waste facilities, 97–98, 153; Yang on, 160n72
environmental law, 21, 153
environmental policymaking, US, 14, 15, 18
Environmental Protection Agency (EPA), 26, 27, 45–46
EPA. *See* Environmental Protection Agency
equality: emissions reduction and, 52, 69; immigration restrictions and, 266. *See also* egalitarians

equity weighting, within cost-benefit analysis, 139n21
"Ethics, Public Policy, and Global Warming" (Jamieson), 53n10
ethics of migration, Stanczyk on, 248
Europe: decarbonization in, 169; temperature in, 191n5
European Union, 30, 278
existentialism: on freedom, 105, 106; of Sartre, 105
expert knowledge: climate policy and, 71–74; political imagination limited by, 66, 68; politics of, 68; public involvement influencing, 73–74; trust in, 71–72
externalities, 83, 145–48, 154; Battistoni on, 45, 141; capitalism and, 111–12, 129, 137; climate change as, 149–52; Condon on, 119n66; of factory smoke, 84–85, 87; history of, 80–81, 84–90; inequality exacerbated by, 82, 125; internalization of, 79; knowledge limitations about, 101; morals of, 88–90; Nozick on, 116n39; Pigou and, 84–86, 141, 145; Pigouvian tax on, 129, 130; politics of, 80, 81, 112; of prices, 102; regulation of, 130, 131; of Scope 1 emissions, 45–46; standard economic environmental policy on, 136; state intervention for, 86, 87–88; uniform disincentives for, 125–26
ExxonMobil, 150–51
Eyal, Nir, 221n15

factory smoke: externality of, 84–85, 87; Pigou on, 84–85, 87, 115n30, 148
failure to act on one's judgment (*akrasia*), 123n123
fairness, 226–27, 228–29, 230, 240
Federal Reserve, 131
Finley, Aaron, 227
Fleurbaey, Marc, 273
Floyd, George, 160n60
Foote, Eunice, 151
forest fires, 165, 185

forever chemicals: harm of, 152; 3M producing, 152, 153, 154
Forster, Piers M., 191n6
fossil fuel industry, 98, 151; Canadian projects in, 169, 187–88; on carbon taxes, 110–11
fossil fuels: harms from, 207; 1.5 degrees Celsius warming ceasing, 171; power plants run on, 21; Tribe representing, 49–50, 53n4
401(k), 148
freedom: of capitalists, 108–9; Hayek on, 105–6, 108, 119n73; price, 107; responsibility relationship with, 112; Sartre on, 105; situated, 7, 252, 254; "Ways Not To Think About Plastic Trees" on, 64n134. *See also* market freedom
Frick, Johann, 285
Friedman, Milton, 86, 94; in *New York Times Magazine*, 152–53; on pollution rights, 92, 112; on state intervention, 116n41
fundamentalism, climate change: Baker on, 253; IRA and, 254
future generations, 1; responsibility toward, 2, 19, 203, 204; utilitarianism on, 177. *See also* intergenerational injustice; intergenerational justice

GDP. *See* gross domestic product
Gell-Mann, Murray, 52n1
gender equality, emissions reduction from, 43, 72
General Motors: Campaign to Make General Motors Responsible, 152–53; Coase on, 143
generous sustainability, 268
geoengineering, solar, 188–89
Germany, fossil fuel industry projects in, 187
GHGs. *See* greenhouse gas emissions
global emissions, 172
global emitter groups, geographical breakdown of, 37, *37*
global energy transition, 168

Index

Global South, population control in, 7
Global Stocktake, under Paris Agreement, 233–34
Goldman Sachs, 32
The Good Place (TV show), 157n28
Gourevitch, Alex, 94
governments: intergenerational justice enacted by, 185; market domination influencing, 108. *See also* state intervention
Great Acceleration, 14–15, 81
green growth, 67, 69
greenhouse gas emissions (GHGs): economic inequality relationship with, 41; economy decoupled from, 48; from human development, 279, 280; individual, 34–35, 36; inequality relationship with, 39–40, 61n95; from meat consumption, 174; offsets for, 29, 30, 33; responsibility for, 27–28; from states, 230; uniform disincentives for, 127; US population control and, 259n36
Green New Deal, 48, 67, 243, 253–54, 263n73; inequality and emissions reduction relationship with, 69; technological challenge and moral difficulty of, 166
green technology solutions, 150, 159n50; for Green New Deal, 166; Stanczyk on, 243; techno-optimist, 245, 254
gross domestic product (GDP): global temperature relationship with, 23–24, *24*, 25, 49, 63n123, 175; population growth and carbon emissions relationship with, 172, 192n11; social welfare relationship with, 175

Hale, Robert, 157n25
Haraway, Donna, 258n30
Hardin, Garrett: on lifeboat ethic, 280; "Tragedy of the Commons" by, 154, 161n80
harm, 99, 229–30; from aviation emissions, 205–7; avoidance, 225, 226,

228; Battistoni on, 158n41; conditional model on, 282–84; disjunctive wrongs on, 183, 224; externality distribution of, 86, 125, 141; of forever chemicals, 152; from fossil fuels, 207; intergenerational justice relationship with, 227–28; nonconsequentialism on, 206, 207–8, 209–10; population growth causing, 182–83; public, 118n56; racial power dynamics enabling, 153; responsibility for, 136
Hayek, Friedrich, 86; on capitalist market, 101–2, 103, 104, 105, 108; on freedom, 105–6, 108, 119n73; on labor, 119n73
Hayes, Denis, 152
HCFC-22. *See* hydrochlorofluorocarbon-22
Hemel, Daniel, 27
HFC-23. *See* hydrofluorocarbon-23
high-income states: emissions of, 272, 275, 281–82, 284, 286, 292n66; human development supported by, 279, 280, 287; immigration restrictions to, 264, 265, 266, 269, 270, 282, 285–86; immigration to, 271–73, 277, 281, 283–84; mitigation of, 284, 285, 286; population growth in, 274–75
household emission levels: within China, 39; elite persons, 37, 39, 42, 59n72; US, 37, *38*
housing market, US, 188
"How Quickly Should the World Reduce Its Greenhouse Gas Emissions?" (Budolfson, McPherson and Plunkett), 194n27
Huber, Matt, 250–51
human development, 265; climate-friendly, 280–82; GHGs from, 279, 280; high-income states supporting, 279, 280, 287; immigration and, 278–82; immigration restrictions and, 278–80, 281–82; population growth relationship with, 248
human right, immigration as, 277

human self-understanding: with immigration, 277–78; with procreation, 276
hydrochlorofluorocarbon-22 (HCFC-22), 29
hydrofluorocarbon-23 (HFC-23), 29, 32

immigration, 291n51; backlash to, 265, 268–69; climate policy on, 248; conditional model of, 282–84; emissions from, 275; to high-income states, 271–73, 277, 281, 283–84; human development and, 278–82; mitigation relationship with, 264, 273; politics impacted by, 168, 265; population growth and, 271–72, 273; special status of, 275–78, 287
immigration restrictions, 8, 44, 292n68; Cafaro and Staples arguments on, 266, 267, 268, 269, 279, 288n9; climate vulnerability from, 273; conditional model on, 283, 284; ecological argument for, 268, 269–70, 289n23; egalitarians on, 264, 265–66, 269, 270, 282, 285, 286; environmental argument for, 264–70, 276, 278–81, 282–83, 285–87, 288n9; global argument for, 267, 268–69; to high-income states, 264, 265, 266, 269, 270, 282, 285–86; human development and, 278–80, 281–82; justification of, 282–86; local argument for, 267, 268, 269; nativist, 265; non-ideal reason for, 285; to US, 266, 267
"Indicators of Global Climate Change 2022" (Forster, Smith, Christopher J. and Walsh), 191n6
inequality, 47; carbon, 36–37; carbon emissions relationship with, 40, 72, 244; of class society, 93; degrowth relationship with, 69; emissions gap from, 59n76; externalities exacerbating, 82, 125; gender, 43, 72; GHG relationship with, 39–40, 61n95; mitigation relationship with, 41, 42–43, 44, 51–52; standard economic environmental policy ignoring, 125–28, 136; uniform disincentives exacerbating, 127, 128. *See also* economic inequality; environmental justice; structural domination
inequality reduction, 3; as climate policy, 67–71; mitigation through, 66–67, 68–69
Inflation Reduction Act (IRA) (2022), 15, 110, 149–50, 243; anti-climate-fundamentalist components of, 254; corporate tax credits in, 263n74; industrial policy focus of, 251; limitations of, 255
instrumentality, toward animals, 186
instrumental rationality, 17
intent, responsibility and, 142, 152–53
Interest Theory of rights, 220n6
intergenerational injustice, 185; of can-kicking, 6, 167–68, 184, 204; of emissions, 180–81; of luxury consumption, 181, 183, 203–4, 205, 209; of population growth, 183
intergenerational justice, 5–6, 219n3; Arrow on, 195n29; Cafaro on, 268; carbon budget allocations for, 7; of Delayed-Action Bomb case, 216–17, 220n11; fairness within, 228–29; governments enacting, 185; harm relationship with, 227–28; luxury emissions reduced by, 184–85; moral responsibility for, 223–24; nonconsequentialist view of, 179, 183–84, 201, 206–8, 209, 211–14; Non-Identity Problem for, 178, 180, 183, 184, 213, 220n12, 223–24, 229, 249; person-regarding view of, 179, 183–84, 201, 211–13, 214, 229; responsibility for, 201–5; restrictions for, 167, 181–82, 184; Stanczyk on, 199–205, 209–12, 215, 226–27, 242, 249–50; UNFCCC calling for, 222; utilitarianism on, 177, 249. *See also* future generations
intergenerational Lockean proviso, 211, 220n8
intergenerational policy issues, 23

Index

Intergovernmental Panel on Climate Change (IPCC): on AMOC, 191n5; on carbon capture, 245; on carbon removal, 245, 257n22; carbon tax proposal of, 91, 109–10; "Demand, Services, and Social Aspects of Mitigation" by, 243; on GDP and global temperature, 175; on human development pathways, 280
international justice, 181
IPAT. *See* environmental impact, population, affluence, and technological development
IPCC. *See* Intergovernmental Panel on Climate Change
IQ, person-regarding, 219n2
IRA. *See* Inflation Reduction Act

Jamieson, Dale, 53n10
Jauernig, Anja, 221n15
Jensen, Michael, 145, 147
Jerneck, Max, 150
Jevons paradox, 69
Justice40 Initiative, 15

Kamm, F. M., 6
Kapp, K. W.: on climate policy, 110; on politics, 91, 92; on social costs, 91, 92, 93, 157n25; *The Social Costs of Private Enterprise* by, 91
kidnapper, justifications of, 285
Krause, Sharon, 111, 124n132
Kyoto Protocol, 28, 222–23
Kysar, Douglas A., 81; on inequality reduction and mitigation, 68; on welfare economics and politics, 2–3

labor: under capitalism, 94–95; Hayek on, 119n73; pollution rights sale compared to, 97; price relationship with, 107; social costs compared to, 96
Latour, Bruno, 154
law, climate change, 1, 253, 261n64
Law and Political Economy movement, on economic inequality, 252

legal orders, 263n80
Lichtenberg, Judith, 99, 100
lifeboat ethic, 280
liquefied natural gas (LNG), 196n41
lithium, 192n9
LNG. *See* liquefied natural gas
luxury consumption, 199; of elite persons, 202; intergenerational injustice of, 181, 183, 203–4, 205, 209; reproductive rights compared to, 209–10; Stanczyk on, 246
luxury emissions: decent-living emissions compared to, 244, 245; of elite persons, 245, 258n31; intergenerational justice reducing, 184–85; policy intervention with, 38; reduction of, 42–43, 68; subsistence emissions compared to, 35, 244

MacGilvray, Eric, 83, 102, 103
Malhi, Yadvinder Singh, 30
Manatee Bay (Florida), 185
Mandeville, Bernard, 85, 103, 109
marginalist revolution, 84
market dependence, 81
market domination: governments influenced by, 108; nonresponsibility and, 83, 98–104; Pettit on, 120n77. *See also* unfreedom
market failure, 79, 81, 85, 88. *See also* externalities
market freedom: of externalities, 101; inadequacy of, 105, 106, 108; nonresponsibility of, 83, 98–104
market participants, responsibility of, 122n106
market regulation, 47
markets, capitalist: anticipatory relationships within, 106–7; habitable planet and, 149; Hale on, 157n25; Hayek on, 101–2, 103, 104, 105, 108; of pollution rights, 89, 90; responsibility within, 90; seriality of, 105–9. *See also* consumer choices
Markey, Edward J., 263n72
Marshall, Alfred, 84

Marx, Karl: on capitalism, 80; on commodity fetishism, 100
McPherson, Tristram, 194n27
Means, Gardiner C., 156n15
meat consumption, 173–74, 202, 256n17
Meckling, William, 145, 147
methodological question, 2
migration project, 277
migration-related emissions, 265
Mildenberger, Matto, 151
Miller, David, 233, 267
mimicking duties, 232
mitigation, 270–75; adaptation trade off with, 188, 189; burden-sharing, 225, 229, 231, 240; can-kicking of, 223, 224, 233, 236, 238; collective, 232–33, 236; economic inequality addressed with, 42, 66; egalitarians on, 264, 285; of high-income states, 284, 285, 286; immigration relationship with, 264, 273; individual, 232–33, 236, 239; through inequality reduction, 66–67, 68–69; inequality relationship with, 41, 42–43, 44, 51–52; political action for, 232; slack-taking for, 233; stabilization wedge approach to, 173, 174, 193n20, 255n5; stock-taking of, 233–36; US, 234; from women parliamentary representation, 43
models: dangers and limitations of, 15–17, 22–25, 49; energy equity and emissions related through, 42; inputs and contestability of, 2–3, 21, 27; integrated assessment models, 19–25, 46, 175, 186
Moellendorf, Darrel, 274
Montana Supreme Court, 32
moral agency, 99; within capitalist market, 108; knowledge limiting, 100
moral responsibility, 168, 190; of consumer, 142; for environmental degradation, 266–68; for intergenerational justice, 223–24; of population growth, 182–83

morals: of degrowth, 72–73; elite persons failing, 261n62; of externalities, 88–90; Green New Deal difficulties with, 166
Moyn, Samuel, 252
Musk, Elon, 147

National Academies of Sciences, Engineering, and Medicine committee, 197n51
National Emissions Trading Scheme (China), 130
nationally determined contributions (NDCs), 231, 237–38
National Science Foundation (US), 52n1
nativist immigration restrictions, 265
nativist public policy, 248
natural capital, 46
natural gas, US producing, 169
Nature, 69, 70
Nature Climate Change, 38–39
"The Nature of the Firm" (Coase) (1937), 145; Chicago School on, 142, 143, 144, 149; on corporations, 143; on employer and employee relationship, 156n13
NDCs. *See* nationally determined contributions
Negative Emissions Technologies and Reliable Sequestration (National Academies of Sciences, Engineering, and Medicine committee) (2019), 197n51
neoclassical economics, 42, 44–45
neoliberal climate thinking, 15, 16, 73
neo-Malthusianism, 259n37
New Deal, 153
New York Times Magazine, Friedman in, 152–53
nexus of contracts theory, of corporations, 144–45, 146, 156n19
No-Difference View, 214
non-anthropocentric ethic, of Cafaro and Staples, 289n21
nonconsequentialism, 5, 6; on harm, 206, 207–8, 209–10; of

intergenerational justice, 179, 183–84, 201, 206–8, 209, 211–14; on Non-Identity Problem, 216; of population growth, 179–80; on rights, 207–8; Stanczyk on, 199, 205, 208–9, 211–12
non-human beings, 268
Non-Identity Problem, 5, 6, 113n12, 261n53; for intergenerational justice, 178, 180, 183, 184, 213, 220n12, 223–24, 229, 249; nonconsequentialism on, 216; Parfit on, 240n3; person-regarding relationship with, 214–15; rights relating to, 218; Stanczyk on, 200, 249
non-person-affecting approach, 201
non-reformist reforms, 7, 252–53
nonresponsibility, market domination and market freedom and, 83, 98–104
Nordhaus, William, 19, 49
North Korea, 130
Nozick, Robert: *Anarchy, State, and Utopia* by, 86; on externality, 116n39

Obama, Barack, 21; executive actions of, 19; US cost-benefit analysis under, 18
Obama administration: can-kicking under, 237; Clean Power Plan of, 49, 50, 238; NDC under, 237, 238
Obama task force, 23; on discounted future cost-benefit analysis, 19, 22; on social cost of carbon, 26, 27; Trump administration disbanding, 25; US cost-benefit analysis of, 22
Ocasio-Cortez, Alexandria, 263nn72–73
Occidental Petroleum, 198n52
ocean temperatures, 185
O'Connor, James, 118n63
offsets: for GHGs, 29, 30, 33; for HFC-23, 29; politics of, 28–29, 31
oil: Occidental Petroleum, 198n52; from Russia and Canada, 169; US and UK producing, 187
1.5 degrees Celsius warming, 23, 37–38, 46, 48, 171
Oppenheimer, Michael, 273

optimal pollution, of Paris Agreement, 228
Otsuka, Michael, 221n15
Our Children's Trust, 32

Pacala, Stephen, 173, 174, 193n20, 255n5
Pamuk, Zeynep, 3
Parfit, Derek, 82, 240n3
Paris Agreement, 55n27, 222–23; Global Stocktake under, 233–34; on global temperature, 228; NDCs for, 231, 237–38
Peabody Energy, Tribe representing, 49, 50
person-regarding, 202; on eggs and people, 219n2; intergenerational justice, 179, 183–84, 201, 211–13, 214, 229; Non-Identity Problem relationship with, 214–15; population growth, 179–80; Stanczyk on, 199, 205, 211–12, 214, 219n2
Pettit, Philip, 102–3, 120n77
philosophy, 1
Pigou, Arthur C., 109, 118n56; Coase compared to, 86–90; externality and, 84–86, 141, 145; on factory smoke, 84–85, 87, 115n30, 148; on politics, 89; on state intervention, 149; on *Wealth of Nations*, 115n33; on welfare economics, 84
Pigouvian tax, 92, 129, 130, 132
plastic age, 152
Plunkett, David, 194n27
Polanyi, Karl, 91
policy feedback, 261n64
policymakers, 3; duties of, 250–55; economist, 126; politically sensitive, 243; scientific disagreement influencing, 70
political action: for mitigation, 232; scientific disagreement influencing, 70–71
political economy: of air pollution regulation, 151; social cost within, 110

political empowerment, of women, 43
political imagination, 44, 67, 72
political imagination, limited: about carbon upsets, 33; collective decision-making excluded by, 31; expert knowledge influencing, 66, 68; Kysar on, 2–3; Pamuk on, 3; uniformity influencing, 73
politically constructed climate targets, 228, 229, 240
political order, 252, 254
political philosophy, 139n19
politicians, Canadian, 169
politics, 1, 261n64; capitalist, 167; of carbon upsets, 32–33, 51, 58n61; climate ethics role in, 7, 8, 243; of collective decision-making, 33; domination in, 97–98, 112, 245; economics relationship with, 80; elite persons influencing, 245; of expert knowledge, 68; of externalities, 80, 81, 112; immigration impacting, 168, 265; Kapp on, 91, 92; Kysar on, 2–3; of market dependence and market failure, 81; of offsets, 28–29, 31; Pigou and Coase on, 89; policy-makers influenced by, 243; state intervention change of, 255; of US cost-benefit analysis, 27
pollution rights, 95; of capitalist markets, 89, 90; Coase on, 91–92, 96, 112; Friedman on, 92, 112; sale of, 97; structural domination of, 96
population control, 154, 167, 185, 260n43; forced sterilization and eugenics for, 195n32; in Global South, 7; horrors of, 248–49; Shultz condemning, 259n37; in US, 259n36
population ethics, 167, 176, 178, 199
population growth: Cafaro and Staples on, 274; consumption growth and, 173–78, 193n21, 220n7, 242, 247–48; economy relationship with, 260n39; emissions relationship with, 174, 192n11, 193n19, 260n41; GDP and carbon emissions relationship with, 172, 192n11; Haraway on, 258n30; in high-income states, 274–75; human development relationship with, 248; immigration and, 271–72, 273; immigration restriction environmental argument invoking, 266; intergenerational injustice of, 183; IPAT formula on, 270–71; nonconsequentialist and person-regarding view of, 179–80; through procreation, 275–76; restrictions on, 182–83; stabilization wedge mitigation approach offsetting, 193n20; Stanczyk on, 220n7, 242, 246–47, 248–49, 252; utilitarian approach to, 176, 177
Posner, Eric, 170
poverty, 274
power-weighted social decision rule, inequality relationship with, 40
prices, 123n122; Chicago School on, 146; externalities of, 102; freedom limitations of, 107; knowledge limitations of, 101; of US housing market, 188
"The Problem of Social Cost" (Coase) (1960), 86–88, 141, 145–46
procreation, special status of, 275–76
professionalization, 73
promotional duties, 231, 232
public harms, 118n56
public philosophy, 225
public policy, nativist, 248

race, structural domination through, 97–98
Rawls, John, 118n56
Reasons and Persons (Parfit), 240n3
recession, global (2009), 19
reforestation, carbon emission relationship with, 174
regulation, 16, 34, 56n37, 230; of air pollution, 129, 151; within capitalism, 4, 128–29, 130, 131, 134, 135, 137; of consumption, 244; of European household energy use, 42; excess democracy contrasted

Index

with, 139n19; of externalities, 130, 131; market, 47; market domination influencing, 108; social cost of carbon influencing, 19, 21; state intervention for, 86, 87; Summers on, 93; US climate policy, 15. *See also* well-regulated capitalism
regulatory cost-benefit analysis, 13
regulatory US cost-benefit analysis, 18, 21–22, 25–26, 27
renewable energies, 172
reproductive justice, 247; Roberts, Dorothy, on, 259n37; in US, 248
reproductive rights, 209–10, 260n42
Repugnant Conclusion, population ethics and, 176, 178, 199
"Research Needs Concerning the Incorporation of Human Values into Environmental Decision Making.," 52n1
responsibility, 17–18; Battistoni on, 142; Cafaro and Staples on, 279; for can-kicking, 225, 230–33; for capitalism, 133; within capitalist markets, 90; of elite persons, 133, 134, 154–55, 253; freedom relationship with, 112; to future generations, 2, 19, 203, 204; for GHGs, 27–28; for harm, 136; individual, 34, 133, 135–36; intent and, 142, 152–53; for intergenerational justice, 201–5; of market participants, 122n106; question, 2, 5. *See also* consumer responsibility; corporate responsibility; intergenerational injustice; nonresponsibility
rights, 221n14; consequentialism, 205–15; nonconsequentialism on, 207–8; Non-Identity Problem relating to, 218; reproductive, 209–10, 260n42; Stanczyk on, 208, 209, 216, 220n6; victim-focused, 217
Roberts, Dorothy, 259n37
Roberts, William Clare, 104
Robeyns, Ingrid, 275–76
Russia: oil produced in, 169; oligarch from, 38

Sandel, Michael, 89
Sartre, Jean-Paul, 83; *Critique of Dialectical Reason* by, 106; on freedom, 105; on seriality, 105, 106
Satz, Debra, 93
science court, 3, 71
scientific consensus, scientific disagreements compared to, 70
scientific disagreements: assumptions and values leading to, 70–71; political imagination influenced by, 67, 72; public scrutiny of, 67–68, 71
scientific research, public involvement in, 73–74
Scope 1 emissions, externalities of, 45–46
second contradiction, 118n63
seriality: of capitalist markets, 105–9; Sartre on, 105, 106
shareholder primacy, 158n37; for board of directors, 147–48; of corporations, 5, 142
Sherman Act, 146–47
Shue, Henry: on subsistence emissions and luxury emissions, 35; on technological development, 274, 290n35
Shultz, Susanne, 259n37
situated freedom, 7, 252, 254
slack-taking, can-kicking compared to, 233
slavery, 218
slave sugar, 100
Smith, Adam, 80, 85, 109, 115n33
Smith, Christopher J., 191n6
social cost of carbon, 44; Biden administration on, 26; EPA on, 26, 27, 45–46; regulation influenced by, 19, 21; Trump administration disavowing, 25; uncertainty of, 54n20; US climate policy on, 26–27; US cost-benefit analysis on, 19–20, 27
social costs: Battistoni on, 157n25; Coase on, 86–88, 90, 110, 141; Kapp on, 91, 92, 93, 157n25; labor compared to, 96; within political economy, 110; structural domination of, 90–98

The Social Costs of Private Enterprise
(Kapp) (1950), 91
social domination, 83
social justice, 181
social welfare, 20; of animals, 196n36;
  in China, 131; externalities ignoring,
  85, 89; GDP relationship with, 175
Socolow, Robert, 52n1, 173, 174,
  193n20, 255n5
solar geoengineering, 188–89
solar industry, US, 150
Southern Company, decarbonization
  of, 246
Soviet Union, 130
speaker-relative justifications, 285
stabilization wedge mitigation
  approach: of Pacala and Socolow,
  173, 174, 193n20, 255n5; population
  growth offset with, 193n20
Stanczyk, Lucas, 5–6; on can-kicking,
  226; on consumption growth,
  220n7, 242, 243–44, 252; disjunctive wrongs applied by, 224; on
  elite persons, 202, 245, 251, 255; on
  intergenerational justice, 199–205,
  209–12, 215, 226–27, 242, 249–50;
  on luxury consumption, 246; on
  nonconsequentialism, 199, 205,
  208–9, 211–12; on Non-Identity
  Problem, 200, 249; on person-
  regarding, 199, 205, 211–12, 214,
  219n2; on population growth,
  220n7, 242, 246–47, 248–49, 252;
  on rights, 208, 209, 216, 220n6
standard economic environmental
  policy: Battistoni critiquing, 132–37;
  of economists, 126; environmental
  justice advocates against, 127; externalities handled by, 136; inequalities
  ignored by, 125–28, 136
Staples, Winthrop, III, 288n20;
  immigration restriction arguments
  by, 266, 267, 268, 269, 288n9; non-
  anthropocentric ethic of, 289n21; on
  population growth and consumption
  growth, 274; on responsibility, 279

state intervention: Coase on, 148–49;
  for externalities, 86, 87–88; Friedman on, 116n41; of Inflation Reduction Act, 15, 110; political change
  for, 255; for regulation, 86, 87
states: can-kicking of, 231–32; carbon
  budget stock-taking for, 235; GHGs
  from, 230; post-colonial status of, 98;
  under UNFCCC, 230–31
Stern, Nicholas, 79
Stigler, George, 88
Stilz, Anna: on conditional model, 282,
  283; on immigration restrictions, 285
stock-taking, 240; of carbon budgets,
  235–36; in climate ethics, 225; of
  mitigation, 233–36
Stokes, Leah, 151
Strontium-90, 152
structural domination, 83; of capitalism, 94–96; of climate vulnerability,
  95–96; of externalities, 111–12;
  through nation and post-colonial
  status, 98; of pollution rights, 96;
  through race, 97–98; of social costs
  and class society, 90–98; Young on,
  111, 161n75
Sub-Saharan Africa, population growth
  in, 260n41
subsistence emissions, luxury emissions
  compared to, 35, 244
sulfur dioxide, 151
sulfur pollution, 171
Summers, Larry, 92; on regulation, 93;
  on structural domination, 97
Sunstein, Cass, 18
supply chains, 157n27
Supreme Court, US, 15
Switzerland, on carbon removal, 246

Tank, Lukas, 244
taxes, 62n108; IRA credits for, 263n74;
  Pigouvian, 92, 129, 130, 132; uniform, 126. *See also* carbon taxes
technological development: disagreements over, 274; of high-income
  states, 273; IPAT formula on, 270–71;

Shue on, 274, 290n35. *See also* green technology solutions
techno-optimist green technology solutions, 245, 254
temperature, global, 79, 191n5; adaptation limit relationship with, 23–25; GDP relationship with, 23–24, *24*, 25, 49, 63n123, 175; increase, 165–66; of ocean, 185; 1.5 degrees Celsius warming, 23, 37–38, 46, 48, 171; Paris Agreement on, 228. *See also* 2 degrees Celsius warming
A Theory of Justice (Rawls), 118n56
3M, forever chemicals produced by, 152, 153, 154
toxic waste markets, 93
"Tragedy of the Commons" (Hardin), 154, 161n80
transaction costs, Chicago School on, 143, 151–52
transportation infrastructure, 236
Tribe, Laurence H., 51; Clean Power Plan opposed by, 49, 50; on discounted future cost-benefit analysis, 13; fossil fuel interests represented by, 49–50, 53n4. *See also* "Ways Not To Think About Plastic Trees"
Trump, Donald, 18
Trump administration, 20; can-kicking under, 237; US cost-benefit analysis under, 25–26, 27
Tuvalu, 38
Twitter, 147
2 degrees Celsius warming: carbon budget for, 171–72; global economy causing, 173

UK. *See* United Kingdom
UN. *See* United Nations
uncertainty, 1, 185, 188–90; of climate tipping points, 250; of energy transition, 168; probability distributions for, 186–87; scientific disagreement revealing, 70, 71; of social cost of carbon, 54n20; stock-taking reflecting, 235

UNFCCC. *See* United Nations Framework Convention on Climate Change
unfreedom: within capitalism, 98, 108–9; of formerly colonized peoples, 4. *See also* market domination
uniform disincentives: distributive justice ignored by, 127; for externalities, 125–26; inequalities exacerbated by, 127, 128
United Kingdom (UK), 187
United Nations Framework Convention on Climate Change (UNFCCC): for intergenerational justice, 222; Kyoto Protocol, 28, 222–23; states under, 230–31; US presidential administrations on, 237. *See also* Paris Agreement
United Nations (UN), 36, 174
United States (US): Acid Rain Trading program in, 151; carbon budget for, 234, 235; carbon emissions in, 234, 247; carbon taxes in, 126; Clean Air Act, 15, 26, 50; Clean Power Plan, 49, 50, 238; Constitution, 50–51; environmental policymaking in, 14, 15, 18; geoengineering in, 189; household emission levels in, 37, *38*; housing market in, 188; immigration restrictions to, 266, 267; National Science Foundation of, 52n1; natural gas produced in, 169; NDC of, 237–38; oil production in, 187; population control in, 259n36; reproductive justice in, 248; solar industry in, 150; Supreme Court, 15; Trump presidencies of, 18. *See also* climate policy, US; Inflation Reduction Act; Obama, Barack
US Congress, 251
US House Resolution, Markey and Ocasio-Cortez introducing, 263n72
utilitarianism, 195n31, 195n35; consequentialist, 199–200; on emissions reduction, 176; on intergenerational justice, 177, 249; on population growth, 176, 177

Vanderheiden, Steve, 6–7
Veblen effects, 40
victim-focused rights, 217

Walsh, Tristram, 191n6
wars, over resources, 187
waste facilities, environmental justice on, 97–98, 153
"Ways Not To Think About Plastic Trees" (Tribe) (1974), 17; on endogenous cultural influence, 16, 50; on freedom, 64n134; from "Research Needs Concerning the Incorporation of Human Values into Environmental Decision Making.," 52n1; on welfare economics, 13–14, 48–49
Wealth of Nations (Smith, Adam), 115n33
weather, domination of, 124n132, 135
Weisbach, David, 170
welfare economics, 16, 17; on carbon emissions, 51; critiques of, 66, 67; Kysar on, 2–3; Pigou on, 84; "Ways Not To Think About Plastic Trees" on, 13–14, 48–49
well-regulated capitalism, 4, 128–29, 135, 137; democracy relationship with, 139n19; domination avoided under, 134
Welton, Shelley, 7
Wenar, Leif, 101
Who Killed the Electric Car? (2006), 150
women parliamentary representation, 43
World Bank memo, of Summers, 93

Yang, Tseming: on environmental justice, 160n72; on environmental law, 153
Young, Iris Marion, 82; on corporations, 153; on structural domination, 111, 161n75

zero-carbon energy technology, carbon budgets informed by, 235

www.ingramcontent.com/pod-product-compliance
Lightning Source LLC
LaVergne TN
LVHW041954060526
838200LV00002B/15